BIOTECHNOLOGY

An Industry Comes of Age

by Steve Olson
for the Academy Industry Program

National Academy of Sciences
National Academy of Engineering
Institute of Medicine

NATIONAL ACADEMY PRESS
Washington, D.C.   1986

National Academy Press   **2101 Constitution Avenue, NW**   **Washington, DC 20418**

This book is based on a symposium sponsored by the Academy Industry Program, a joint project of the National Academy of Sciences, the National Academy of Engineering, and the Institute of Medicine. It has been reviewed according to procedures approved by a Report Review Committee consisting of members of the two Academies and the Institute of Medicine.

The National Academy of Sciences (NAS) is a private, self-perpetuating society of distinguished scholars in scientific and engineering research, dedicated to the furtherance of science and technology and their use for the general welfare. Under the authority of its congressional charter of 1863, the Academy has a working mandate that calls upon it to advise the federal government on scientific and technical matters. The Academy carries out this mandate primarily through the National Research Council, which it jointly administers with the National Academy of Engineering and the Institute of Medicine.

The National Academy of Engineering (NAE) was established in 1964, under the charter of the NAS, as a parallel organization of distinguished engineers, autonomous in its administration and in the selection of members, sharing with the NAS its responsibilities for advising the federal government.

The Institute of Medicine was chartered in 1970 by the National Academy of Sciences to enlist distinguished members of appropriate professions in the examination of policy matters pertaining to the health of the public. In this, the Institute acts under both the Academy's 1863 congressional charter responsibility to be an adviser to the federal government, and its own initiative in identifying issues of medical care, research, and education.

Library of Congress Cataloging-in-Publication Data

Main entry under title:

Biotechnology: an industry comes of age.

Includes index.
Written and supplemented by additional research materials by science writer Steve Olson from transcripts of a conference held in Washington, D.C., Feb. 27-28, 1985, and sponsored by the Academy Industry Program of the National Academy of Sciences, National Academy of Engineering, and Institute of Medicine.
1. Biotechnology—Congresses.   2. Genetic engineering
—Congresses.   I. Olson, Steve, 1956- .
II. Academy Industry Program (National Research
Council (U.S.)) III. National Academy of Engineering.
IV. Institute of Medicine (U.S.)
TP248.14.B56   1986   660'.6   85-28442
ISBN 0-309-03631-3

Printed in the United States of America

Preface

ON FEBRUARY 24-27, 1975, in a converted chapel on the grounds of the Asilomar Conference Center in Pacific Grove, California, some 140 scientists, several lawyers, and a dozen or so journalists gathered to discuss a profound new development in molecular biology. A few years earlier, researchers had discovered how to isolate and recombine specific segments of DNA, making it possible for the first time to directly manipulate the molecule that gives rise to life. But the new techniques also seemed to pose potential risks—risks that the scientists found impossible to quantify but impossible to ignore. The Asilomar conference, which was sponsored by the National Academy of Sciences, marked an attempt to anticipate and minimize those risks before the research got under way.

Just two years later, in March 1977, the National Academy of Sciences sponsored another meeting on the techniques of genetic engineering, this one in the auditorium of the academy's headquarters in Washington, D.C. In this open forum setting, scientists and individuals from across the country gathered to freely discuss the benefits and risks of the new techniques. But coming as it did at the height of public concern over recombinant DNA research, the forum generated strong emotions. Participants from an overflow audience repeatedly rose to call for drastic restrictions or a halt to the research. Amidst chants of "We will not be cloned," the proceedings frequently came close to breaking down completely.

There were no such pyrotechnics at the most recent academy conference on genetic engineering, held on February 27-28, 1985, to mark the tenth anniversary of the Asilomar conference. Sponsored by the Academy Industry Program of the National Academy of Sciences, the National Academy of Engineering, and the Institute of Medicine, the conference offered a timely demonstration of how the passage of just a few years has affected genetic engineering. For one thing, the conferees were drawn from a much broader range of backgrounds than they were at Asilomar. In addition to scientists, lawyers, and journalists, the roughly 300 attendees included corporate chairmen and vice presidents, economists, university administrators, government regulators, a U.S. senator, and Capitol Hill staffers. The topics under discussion had similarly diversified, ranging from patent law to human gene therapy, from analyses of ecological balances to analyses of international competition.

But perhaps the most telling difference between this conference and the previous two involves the perceived level of risks associated with recombinant DNA research. As experience with the new techniques has accumulated, researchers have realized that the risks discussed at Asilomar and at the Academy Forum were either far overestimated or nonexistent. As a result, the guidelines that govern research with recombinant DNA have been relaxed several times, and almost all the experiments now conducted are exempt from the guidelines.

That is not to say that concerns do not still exist. Participants at the 1985 conference devoted considerable time to discussing the risks associated with human gene therapy or the release of genetically engineered organisms into the environment. In fact, the single most discussed topic at the conference was how the government should regulate, yet not inhibit, useful and safe commercial products now emerging from genetic engineering.

But the concerns have matured. No longer is genetic engineering in its infancy, without a base of scientific information and personal experience on which to build. It has entered what might be characterized as a vigorous childhood, generating its first products and displaying flashes of the virtuosity to come.

The emergence of an ambitious and expanding commercial industry from the seedbed of genetic engineering research has been an important aspect of that growth. Today companies are using the new techniques to develop products that could have dramatic effects on medicine, agriculture, energy production, and many other fields, and in the long run the potential impact of the new industry is virtually unlimited. The industry's remarkable growth is reflected in the main title of this book and of the conference on which it is based, *biotechnol-ogy*. Ten years ago that word, like the industry itself, scarcely existed. Today both are firmly established and are making themselves ever more visible.

This book, which was prepared by science writer Steve Olson, is based on the proceedings of the conference supplemented by additional research materials. Individual chapters and the entire book were reviewed by conference participants and by reviewers selected by the National Research Council.

On behalf of the Academy Industry Program, we wish to thank each of the speakers who addressed the symposium and to express our special appreciation to Alexander Rich, Sedgwick Professor of Biophysics at the Massachusetts Institute of Technology, and Richard J. Mahoney, president and chief executive officer of the Monsanto Company, for cochairing the symposium and for their contributions to the design of its program. For the massive task of organizing the conference, we would like to thank Nancy Gardner Hargrave, staff officer with the National Research Council. Finally, we would like to thank those individuals who contributed their expertise by reviewing and commenting on the symposium's program at various stages: Douglas M. Costle, James D. Ebert, Barbara Filner, Alvin G. Lazen, Robert B. Nicholas, Howard Schneiderman, and Maxine F. Singer.

As genetic engineering enters its second full decade, it is poised to once again capture the public's attention. A number of products of biotechnology are approaching the marketplace—the first installments of genetic engineering's long-heralded promise. Difficult but manageable questions remain to be resolved, questions not only of safety but of regulation, policy, and ethics. Indeed, it is a time of great promise in biotechnology, an apt reminder of the situation that faced the conferecs at Asilomar ten years ago.

> Frank Press
> *President*, National Academy of Sciences
>
> Robert M. White
> *President*, National Academy of Engineering
>
> Frederick C. Robbins
> *Former President*, Institute of Medicine

Acknowledgments

THE ACADEMY INDUSTRY PROGRAM was established in 1983 as a mechanism for bringing the intellectual and financial resources of U.S. industry to the work of the National Research Council and for ensuring the strength of institutional ties to the industrial, scientific, and technological communities. Participating companies, numbering over 60, contribute a total of $1 million each year to support studies, seminars, symposia, and other programs on problems of national consequence for which science and technology are central. The program also provides opportunities for corporate leaders to meet with policymakers from the federal government, universities, and other sectors to discuss national issues.

This publication and its dissemination have been made possible by the National Research Council Fund, a pool of private, discretionary, nonfederal funds that is used to support a program of Academy-initiated studies of national issues in which science and technology figure significantly. The NRC Fund consists of contributions from a consortium of private foundations including the Carnegie Corporation of New York, the Charles E. Culpeper Foundation, the William and Flora Hewlett Foundation, the John D. and Catherine T. MacArthur Foundation, the Andrew W. Mellon Foundation, the Rockefeller Foundation, and the Alfred P. Sloan Foundation; the Academy Industry Program, which seeks annual contributions from companies that are concerned with the health of U.S. science and technology and with public policy issues with technological content; and the National Academy of Sciences and the National Academy of Engineering endowments.

Contents

Four recent and important developments in the regulation of biotechnology should be noted:

1. The White House has approved a federal committee (the Biotechnology Science Coordinating Committee) to coordinate the government's handling of biotechnology issues.

2. Genetically engineered plants, seeds, and tissue cultures can now be patented as a result of a recent decision by an appeals board of the U.S. Patent and Trademark Office.

3. Guidelines on human gene therapy research have been approved by the Recombinant DNA Advisory Committee of the National Institutes of Health.

4. The federal government has approved the first two field tests of genetically engineered organisms: (1) to study the effects of the outdoor environment on plants that have been modified to resist disease, and (2) to observe plants sprayed with bacteria altered to prevent frost formation.

12/85

1

Genetic Engineering and Biotechnology: An Overview

HUMAN BEINGS RELY ON THE EARTH's bountiful supply of life for a wide variety of essential substances. We survive by consuming the edible portions of plants and animals, and our clothes and homes are composed at least in part of biologically derived materials. Microorganisms are used to make bread, to convert milk into cheese, and to brew alcoholic beverages. Common substances like vinegar, vitamins, and monosodium glutamate are manufactured using microbial "factories." Antibiotics are extracted from various strains of molds and bacteria.

Over the course of time, human ingenuity has gradually worked to improve these organisms. People have selected plants, animals, and microorganisms with the most useful characteristics from among those found wild in the environment. They have bred individuals from the same or closely related species to produce offspring with new, more desirable combinations of traits. Among the results of this genetic husbandry have been improved varieties of crops and livestock, industrial microbes that are hardier and more efficient, and novel antibiotics.

During the past 15 years, researchers have begun to acquire a new and unprecedented degree of control over the genetic constitution of living things. The techniques of genetic engineering, and in particular recombinant DNA, have made it possible to manipulate genetic material on the smallest possible scale—individual genes. The effect on

National Institutes of Health

The most important influence of genetic engineering has been on basic scientific research. Genetic engineering has made it possible, using tools such as the DNA map displayed here, to read the genetic code of specific organisms—something that was unthought of before the advent of recombinant DNA.

molecular biology, immunology, and other scientific disciplines has been little short of revolutionary. Says Douglas Costle, former administrator of the Environmental Protection Agency, "While it is probably true that physics was the science of the first half of the century, it is almost certain to be molecular biology in what remains of this century and well into the next."

The development of genetic engineering has been a direct result of generous governmental funding for basic biomedical research since World War II, and it is this research that has benefited most immediately from the new techniques. "The impact of this technology has been enormous at the scientific level," says Philip Leder of Harvard Medical School. "Prior to 1973-74, when these experiments began, all that geneticists knew about the existence of genes they inferred from their properties. . . . Recombinant DNA technology changed that in a stroke. In so doing, it altered genetics from a purely inferential science to, at least in part, an analytical, observational science."

In just a decade of work with recombinant DNA, researchers have uncovered a wealth of new information about how DNA is organized in cells and how it functions. They have found that single genes in higher organisms are usually split into separate and distinct segments of DNA. They have learned a great deal about oncogenes—a class of genes involved in the development of cancer. The exact sequences of the genetic material in a number of viruses, bacteria, and human genes have been determined. "We now know an enormous amount about the genome of many different organisms," says Alexander Rich of the Massachusetts Institute of Technology. "A body of data has accumulated that makes it possible, for example, to consider waging an effective war against a new disease, like acquired immune deficiency syndrome, precisely because of this new technology."

But genetic engineering has done more than give researchers the ability to understand the genetic structure of living things; it has also given them the ability to change that structure. It is now possible to move genetic material in a functional form from one organism to another, creating genetic constructs that have never before existed in nature. For instance, the gene that produces a protein in a human cell can be isolated and inserted into a bacterium. That bacterium can then be reproduced or cloned, creating many identical copies of the gene. If the gene can be coaxed to manufacture the same protein in bacteria that it does in humans, large quantities of the protein can be produced for pharmaceutical applications. And bacteria are not the only possible recipients of new genetic material. Functional genes can be inserted into the cells of plants, animals, and even humans.

This capacity of genetic engineering to introduce completely new traits into existing organisms has given rise to a development that few of the technology's founders could have foreseen. "The thing that we most underestimated ten years ago was the enormous potential that this new technology has for developing an entirely new industry, that of biotechnology," says Rich. "It has given rise to an enormous proliferation of biotechnology companies— over 200 of them in this country alone, and the number is growing." These companies are using genetic engineering to create new kinds of drugs, new vaccines, and diagnostic tools that promise the early detection of disease. They are searching for ways to produce food additives and industrial chemicals more economically through biological means. They are creating genetically altered microbes, plants, and animals to be used in agriculture or in the treatment of wastes.

Biotechnology will have its most immediate impact in certain commercial sectors—such as pharmaceuticals and agriculture—in the industrialized nations. But because it directly affects such basic human concerns as food production, health care, and energy availability, it is likely to eventually have worldwide implications. As Leder says, "It is impossible for us to say with confidence that something reasonable *cannot* be done using this technology."

Any technology that deals so directly with the basic processes of life inevitably raises compelling questions. The early debates about the safety of recombinant DNA research have quieted, but new issues have taken their place. Will the release of genetically engineered organisms into the environment pose threats to human health or to natural ecosystems? How should the ability to alter the genetic makeup of human beings be managed? Is new legislation necessary to regulate the products that are likely to be manufactured with genetic engineering? Should the U.S. government be encouraging the development of the American biotechnology industry in light of the considerable competition expected from biotechnology companies abroad?

These and other difficult questions are being asked with a special urgency. Biotechnology is growing so quickly, and its ultimate influence is so wide-ranging, that it is straining the capacity of public and private institutions to deal with it. "We are running out of time," explains Senator Albert Gore, Jr., "in the sense that the technology is developing so rapidly that we are going to have to make some tentative decisions without the base of understanding that a democracy requires for subtle and difficult decisions. Requests for field tests of genetically engineered organisms are already beginning to be made, as companies proceed with their research programs. The first authorized human gene therapy experiments are expected to be conducted later this year. Both of these facts underscore how important it is to develop a coherent set of scientific and ethical guidelines to help us evaluate the implications of this technology."

The Molecular and Microbial Products of Biotechnology

Most of the products being developed in biotechnology fall into one of two very broad categories: chemical substances that can be made using genetically engineered organisms, and genetically engineered organisms themselves.

Included in the first category are the wide variety of compounds that have drawn the attention of pharmaceutical manufacturers. Genetically engineered microorganisms can be used to produce hormones like insulin and growth hormone, other biological response modifiers such as interferons and neuropeptides, blood products like clotting and antishock factors, vaccines against previously unpreventable diseases, new antibiotics, and many other kinds of biologically active molecules. In addition, the availability of large quantities of these previously scarce molecules enables researchers to learn more about their function in the body, which will result in new therapeutic agents.

The ability of genetically engineered microorganisms to produce valuable chemical compounds will also lead to applications in many other industries, including the food processing, chemicals, and energy industries. Among the numerous substances whose production could be affected by biotechnology are alcohol, enzymes, amino acids, vitamins, high-grade oils, adhesives, and dyes. Biotechnology will also make possible the synthesis of novel chemical compounds in these commercial sectors.

The use of biological processes in industry places special demands on manufacturing. Generally, biological conversions entail a fermentation process. Nutrients and raw materials are supplied to living cells in a reactor vessel; the cells convert the raw materials into products; and the products are withdrawn, separated, and purified. These bioconversions must be carefully monitored and controlled. Indeed, the development of economical fermentation equipment and methods is one of the greatest challenges facing biotechnology today.

But not all genetically engineered microorganisms will be used in fermentation processes. Some are being designed for use in the environment. Many of these will have agricultural applications, but others might be used to degrade wastes or toxic substances, to leach or concentrate minerals from ores, or to increase the extraction of oil from wells.

An important subset of the molecular products of biotechnology are the proteins known as monoclonal antibodies. These are produced not through recombinant DNA techniques but through the fusion of a tumor cell with an antibody-producing white blood cell. The result is a virtually immortal clone of cells producing antibodies that are chemically identical. Monoclonal antibodies have already found a wide range of uses in research, because of their remarkable ability to attach to specific molecular configurations. They are also being used in a number of in vitro diagnostic tests to detect the presence of disease or other conditions. At the same time, investigators are examining their possible uses within the body to expose diseased areas to scanning instruments, to confer passive immunity against disease, or to carry biologically active agents to diseased tissues.

Biotechnology in Agriculture

Many of the products being developed for use in human health care have agricultural analogs. New or cheaper drugs, vaccines, and diagnostics will all cut the toll of disease and lost productivity that continues to be a major concern in agriculture. Furthermore, genetically engineered microorganisms will be used to produce feed additives, growth enhancers, and other compounds that will boost agricultural yields.

But biotechnology has a fundamentally different capability in agriculture. It can potentially be used to change the genetic constitution of microorganisms, plants, and animals to make them more productive, more resistant to disease or environmental stress, or more nutritious. In doing so, biotechnology, like the green revolution before it, could have a dramatic effect on the problems of food production and hunger around the world.

Probably the first application of this type will involve the genetic engineering of microorganisms. Researchers are working to produce microorganisms that will supply plants or animals with essential nutrients, protect them from insects or disease, or provide them with compounds that influence their growth. A central concern of this work is the competitiveness of the genetically engineered microorganisms in agricultural environments, since the microorganisms will generally have to survive and multiply to perform their functions.

The genetic engineering of plants and animals is a far more daunting technical task than the genetic engineering of microorganisms, but this is where the greatest potential benefits lie. Researchers have already succeeded in inserting functional genes into plant cells, in regenerating whole plants that express the gene, and in having the gene passed on to offspring. In this way, they hope to eventually be able to transfer into plants such traits as resistance to pesticides, tolerance to environmental conditions such as salinity or toxic metals, greater nutritive value or productivity, or perhaps even the ability to fix nitrogen from the atmosphere. However, major technical barriers still prohibit the genetic engineering of most of the agriculturally important food crops. For instance, the majority of desirable agricultural traits are likely to arise from the interaction of many different genes, making it difficult to transfer these traits between plants. A major current limitation on research in this area is the paucity of basic biochemical knowledge about plants. To take one example, the genetic origins of almost all agriculturally useful traits are not yet known.

Genes have also been inserted into the sex cells of animals in such a way that they are reproduced in the cells of the mature animal, function in those cells, and are passed on to offspring. For instance, researchers have introduced growth hormone genes into several kinds of agriculturally important animals in an attempt to make the animals grow faster, larger, or leaner. It remains to be seen whether this genetic modification will upset the animals' metabolic balance, causing harmful long-term effects on their health.

Human Gene Therapy

Just as genes can be inserted in a functional form into the cells of animals, so they can be inserted into human cells. There is an important distinction between the genetic engineering of animals and humans, however. For the foreseeable future, genes will be introduced only into limited subsets of a patient's somatic cells. Because the new genes will be reproduced only in that population of cells, they will not be passed on to offspring. Technical difficulties and ethical constraints will rule out the genetic engineering of human sex or germline cells for many years to come.

The first attempts at human gene therapy will involve the insertion of genes into bone marrow cells extracted from patients with severe genetic disorders. The transformed bone marrow cells will be reinserted into the patient's body, where, if the procedure is successful, they will multiply and alleviate the patient's disease. This type of treatment is essentially similar to other kinds of medical procedures, such as transplants, and it raises no new ethical problems.

The technical and ethical problems associated with germline gene therapy are far more formidable. First, the procedures used with animals so damage most of the treated cells that they never develop into live animals. Second, only a fraction of the treated cells that do grow contain the foreign gene. Third, the insertion of a gene can cause severe and often lethal mutations in the cell. Finally, germline gene therapy would alter the genetic pool of the human species, raising fundamental questions about tampering with humanity's genetic heritage.

Ethical considerations are also associated with the use of genetic engineering to enhance a human characteristic, as opposed to replacing a defective gene. In certain cases the issues are clear-cut, as in the condemnation of any attempt to insert a growth hormone gene into an otherwise normal person. But other cases are less well resolved. For instance, it may eventually be possible through human gene therapy to reduce a person's susceptibility to various diseases.

The public's greatest fear about human gene therapy is that it might someday be used to alter such fundamental human attributes as intelligence, character, or physical appearance. However, such traits are undoubtedly shaped by the interplay of many interacting genes with innumerable environmental influences, making it extremely unlikely that they could ever be altered through genetic means. Nevertheless, this fear has helped generate a valuable public dialogue about the capabilities of human gene therapy—a dialogue that should continue as the science evolves.

The Release of Genetically Engineered Organisms into the Environment

Another issue that has generated considerable public discussion in recent years has been the approach of the first field tests of genetically engineered organisms in the environment. It is very difficult to predict exactly what influence a novel organism will exert on an ecosystem, and history is replete with examples of organisms introduced into an environment from elsewhere in the world that had unanticipated, and occasionally devastating, effects. By the same token, conventional breeding techniques have been used throughout history to create new varieties of plants and animals without undue consequences.

To calculate the environmental risk of genetically engineered organisms, five questions must be answered. Will the organism be released into the environment? Will it survive once it is released? Will the organism multiply? Will it move from the place where it is released to a place where it has an effect? And what will that effect be? Furthermore, a genetically engineered organism can sexually or asexually transfer part of its DNA to another organism, which generates a similar string of questions for the organism receiving the DNA.

The chance that a genetically engineered organism will have a detrimental effect on the environment is the product of the five factors listed above. In any given case, the probability that the answer to one or more of these questions will be "yes" is likely to be low, which makes the overall probability of a harmful effect even lower. But it is not zero, and the harmful consequences of a low-probability event could be substantial.

Reducing the uncertainties that surround the effects of genetically engineered organisms on the environment requires additional research focusing on each of the factors that contribute to environmental risk, with the goal of ensuring that the initial field tests are as safe as possible. As with the basic techniques of genetic engineering, it will then be possible to build on a base of experience in expanding the range of environmental uses for genetically engineered organisms.

Governmental Regulation of Biotechnology

The federal government regulates biotechnology from two distinct perspectives. In the area of research, the National Institutes of Health, through its Recombinant DNA Advisory Committee (RAC), has established guidelines that prohibit certain kinds of experiments and set various levels of containment for others. The guidelines apply only to federally funded research, but nongovernmental research institutes and private companies have also adopted them. An increasingly larger portion of the research has become exempt from the guidelines as the level of concern over the risks of recombinant DNA research has fallen during the past decade.

The federal government also regulates biotechnology through the actions of various agencies with authority over emerging products. The Food and Drug Administration (FDA) approves new human drugs and biologics, food additives, medical devices, and some agricultural products and veterinary medicines. The Environmental Protection Agency (EPA) regulates pesticides, hazardous chemicals, and pollutants and plans to oversee the release of certain genetically engineered organisms into the environment. The U.S. Department of Agriculture regulates animal biologics and broad categories of organisms important to agriculture, a jurisdiction that partially overlaps the jurisdictions of the FDA and the EPA. Each of these agencies, in regulating the products of biotechnology, also becomes involved to some extent in overseeing the research and development leading to those products.

In response to apprehensions about such issues as overlapping jurisdictions, the division of responsibility between the RAC and other federal agencies, and the adequacy of existing legislation to ensure the safety of forthcoming applications of biotechnology, the Cabinet Council on Natural Resources and the Environment created the Cabinet Council Working Group on Biotechnology in 1984 under the leadership of the White House Office of Science and Technology Policy. The working group concluded that no new legislation was needed to give federal agencies adequate regulatory authority over the products of biotechnology expected in the immediate future. However, the group did propose that committees similar to the RAC be set up at each of the federal agencies with significant jurisdiction over biotechnology. It also proposed the formation of an interagency coordinating committee on biotechnology, which would lend direction to the science underlying biotechnology's regulation. These proposals have been criticized for setting up additional layers of bureaucracy in the regulatory process and for ignoring the RAC's capacity to handle anticipated regulatory problems.

Industry leaders and government regulators agree that a stable and sound regulatory regime is essential for the continued development of biotechnology. If the public perceives that regulatory agencies are not acting to ensure health and safety, it can move to slow down or halt a technology's development. Public trust could also be fostered through a comprehensive and trustworthy program of public education that clearly lays out both the benefits and the risks of biotechnology.

The New Biotechnology Firms

Two types of firms are pursuing the commercialization of genetic engineering in the United States: small entrepreneurial firms founded almost exclusively since 1976 specifically to capitalize on research developments in genetics, and established multiproduct firms in traditional industrial sectors such as pharmaceuticals, chemicals, energy, agriculture, and food processing. The interactions and complementary attributes of these two types of firms have contributed greatly to the lead in biotechnology that the United States currently enjoys.

The start-up biotechnology firms, of which there are now more than 200, have acquired financing from a variety of sources. Early in their histories they relied heavily on equity investments, research agreements, and licensing contracts with larger firms that wanted a window on the new technology. More recently, these firms have been turning toward other funding mechanisms, such as public stock offerings and R&D limited partnerships, to achieve greater managerial independence and the possibility of larger returns on their investments.

As biotechnology moves beyond research and early product development, the start-up firms will face new challenges. Large established firms are setting up major in-house programs in biotechnology, heightening the already acute competition in the field. To survive, the new firms will eventually have to become profitable through the sale of products. Some firms have pursued this requirement by licensing some or all of their initial products to established companies in exchange for royalties. Others are setting up large-scale production facilities and marketing systems. The success of this latter group, given a product with a market advantage, will depend largely on the availability of further capital to finance scale-up, clinical tests, production, and distribution.

Patents and Trade Secrets in Biotechnology

A prominent concern of all companies involved in biotechnology is the degree of protection they can obtain over the products and processes they develop. In the United States this protection takes two main forms: patents and trade secrecy.

In 1980 the Supreme Court ruled that a genetically engineered microorganism could be patented. Although it remains unclear if higher organisms can be patented under similar provisions, this ruling has cleared the way for a wide variety of patent applications and approvals in biotechnology.

One problem with patents in biotechnology involves the requirement that patented inventions be described in enough detail that they can be reproduced without undue experimentation. Because microorganisms generally cannot be described in such detail, courts have stipulated that this requirement must usually be met by depositing a sample of the microorganism in a culture depository. This gives competitors direct access to the microorganism, increasing the possibility of patent infringement. Ways to restrict access to these deposits without violating the requirements of the patent law are being considered.

If the acquisition or enforcement of a patent appears difficult, a company may rely instead on trade secrecy laws to protect a product or process. In the United States the holder of a trade secret can obtain an injunction or monetary damages in state courts against a party who acquires the secret through improper means. However, there are several drawbacks to trade secrecy laws. For one, they offer no protection against someone who independently discovers the secret, who may then patent it and prohibit the original party from using it. Also, some states are less protective than others of the results of research. Trade secrecy bars scientists from publishing the results of their research in the scientific literature. And the theft of a trade secret is often difficult to prove in court. Finally, it may be necessary to release trade secrets in public forums to demonstrate the safety of a proposed experiment.

University-Industry Relations

Most of the basic techniques that gave rise to biotechnology were originally developed in university laboratories and other research institutes, and biotechnology today remains perched on the leading edge of research. For that reason, industry has a vital interest in establishing and maintaining ties with academic research institutes.

Less well recognized than the benefits to industry are the benefits to universities from university-industry alliances, in addition to the obvious attraction of additional sources of revenue. Such alliances create new challenges for academic science and engineering, place undergraduate and graduate education in new perspectives, increase scientific communication and cooperation, and tie university programs more closely to national and regional needs.

Universities and industry have established a wide variety of cooperative agreements related to biotechnology, including consulting arrangements, industrial associates programs, research contracts, independent research institutes, and private companies affiliated with universities. But at least some of these arrangements involve the possibility of serious conflicts of interest for the researchers and institutions involved. For instance, conflicts may arise over the need for industrial secrecy, the retention of patent rights, or the commercial orientation of research.

Many of these issues were extensively discussed in national forums during the early 1980s, when a number of alliances were being formed in biotechnology. Since then, the debate has become more specific and has moved to the local level as universities and industry gain experience with the first wave of agreements.

Biotechnology in Japan: A Challenge to U.S. Leadership?

The United States currently enjoys a sizable lead in transforming the results of basic biomedical research into commercial products. Other industrialized countries, however, recognizing the economic potential of biotechnology, have adopted national policies to encourage its development. The Japanese government, in particular, has organized research consortia among companies, has sponsored research into biotechnology by industry, and has greatly stepped up its overall funding of biotechnology research. The U.S. government still spends much more money on biotechnology research than does the Japanese government. But the Japanese support for biotechnology is focused largely on applied research, such as the development of fermentation technologies, whereas the U.S. government's support for biotechnology is now overwhelmingly directed at basic research.

The industrial policies of Japan and the United States strongly influence biotechnology in the two countries. If the U.S. government wished to boost the competitiveness of domestic biotechnology firms, it could do so indirectly through changes in these policies. For instance, the tax and investment laws of the United States have been very conducive to the formation of start-up biotechnology firms because of the venture capital they make available to entrepreneurs. There are very few start-up biotechnology companies in the rest of the world— and none in Japan—largely because of the more conservative financial climates abroad.

Japan's regulation of biotechnology is similar to that of the United States, although its regulation of genetic engineering research and of new drugs, biologics, and medical devices is in some ways more restrictive than is U.S. regulation. In the past, Japan has used its strict regulations as nontariff barriers to the import of pharmaceuticals and other products. Japanese laws have been changed to give equal treatment in principle to foreign products, but significant administrative and social barriers to such imports still exist.

The range of patentable subject matter is not quite as broad in Japan as in the United States, and Japan's grace period for filing a patent application after the public release of the patented information is just 6 months, compared with 12 months in the United States. Moreover, patent applications are made public in Japan about 18 months after the filing date, precluding the option of trade secrecy once a decision is made to pursue a patent.

Japan has sought to compensate for deficits in disciplines related to biotechnology by retraining Japanese scientists, engineers, and technicians; by sending researchers abroad to study; and by inducing Japanese nationals working abroad to return to the country. It has also drawn upon its extensive historical experience with fermentation techniques in developing production methods in biotechnology. A number of Japanese researchers are studying biotechnology-related subjects in the United States; the corresponding number of Americans traveling abroad to study biotechnology is very low, even though there are a number of eminent foreign research institutes that could offer valuable training.

Finally, the diffusion of information about developments in genetic engineering and biotechnology is much more extensive in the United States than in Japan. Japanese companies have also purchased a considerable amount of contract research from American biotechnology firms, which gives them access to the state of the art in biotechnology. Both of these factors contribute to what many observers agree is a net transfer of technology from the United States to Japan. This is one of the ways in which the Japanese have been able to mount a strong effort in biotechnology so quickly.

2

The Molecular and Microbial Products of Biotechnology

T HE USE OF BIOTECHNOLOGY in industry often entails a fundamental shift in manufacturing procedures. Biotechnology is based on biological synthesis, usually in water-based solutions at close to room temperature, rather than on chemical synthesis, which often takes place at high temperatures and pressures. This basic attribute of biotechnology gives rise to much of its promise as well as to many of the problems encountered in its large-scale applications.

The molecular products of biotechnology fall into three overlapping categories: new substances that have never before been available, rare substances that have not been widely available, and existing substances that can be made more cheaply through biotechnology. Many of these substances are targeted at human health care: genetically engineered microorganisms can be used to produce hormones, immune regulators, vaccines, blood products, antibodies, antibiotics, and many other biologically active molecules.

Other commercial sectors, such as the food additive and specialty chemicals industries, are also investigating the use of genetically engineered organisms to make new or scarce products or to make

This chapter includes material from the presentations by Philip Leder and William E. Paul at the Symposium on Biotechnology: Creating an Environment for Technological Growth.

14

existing products more cheaply. Enzymes, amino acids, vitamins, high-grade oils, adhesives, and dyes are all examples of substances that could be manufactured through biotechnology. As the science evolves and production costs drop, even some industrial chemicals now made from petroleum and natural gas feedstocks might be produced by microorganisms.

The molecular products of biotechnology are made through fermentation processes, and the design of cost-efficient fermentors and associated production techniques is a major concern in the industry. But biotechnology will also yield genetically engineered microorganisms that have more direct applications. Many of these will be in agriculture, but genetically engineered microorganisms might also be used to decompose sludge at wastewater treatment plants, to leach minerals from low-grade concentrations of ore, or to decrease the viscosity of oil deep underground to allow it to be pumped to the surface. In some of these cases, naturally occurring microorganisms already contribute to these processes. These are possible examples, therefore, of how genetic engineering could be used to expand and improve upon the traditional uses of microorganisms in industry.

The Molecular Machinery of the Cell

The feature of life on earth that makes genetic engineering possible is the universality of the genetic code. Every living organism uses virtually the same system to translate the information contained in its DNA into proteins, the workhorses of biochemistry. It is this common genetic language that enables researchers to reproduce a gene from a human cell, insert it into bacteria, and have those bacteria manufacture the protein encoded by that gene.

Proteins are composed of 20 different, relatively simple molecules known as amino acids, strung together in chains of widely varying lengths. The sequence of amino acids in a protein determines how the amino acid chain will fold, resulting in a characteristic shape that enables a protein to carry out its function. In addition, some proteins consist of two or more amino acid chains bound together; some amino acids are chemically modified once they become part of certain proteins; and some proteins must have other molecules, such as sugars, attached to them before they can function.

By far the largest category of proteins is made up of the enzymes— large, globular proteins that catalyze individual chemical reactions, generally making them occur at least a million times faster than they would in the absence of the enzyme. Other, smaller proteins are hormones, chemical messengers that modify and coordinate the activities of cells. Proteins give bone and skin their tensile strength, and they are involved in the transport and storage of essential molecules within the body. Various proteins produce the movement of muscles, provide immune protection as antibodies, generate and alter nerve impulses, and control growth and differentiation. Clearly, to understand the molecular machinery of the cell, it is necessary to understand the construction and function of proteins.

The basic process by which the information encoded in the genes of an organism's DNA directs the synthesis of proteins was worked out during the 1950s and 1960s. (The books listed in the section Additional Readings at the end of this chapter all describe this process.) But as late as 1970, molecular biologists faced serious difficulties in trying to investigate that process in specific organisms. They had no way of directly manipulating the DNA within higher organisms to determine the details of its structure or function. This problem was exacerbated by the complexity of the DNA in higher organisms, which almost guaranteed that progress would be arduous. "If we took the DNA from a single set of chromosomes from a single human cell and laid it out, it would be about one meter in length," explains Philip Leder of Harvard Medical School. "If we could stretch that one meter into one kilometer, a single gene would be represented in a millimeter's worth of DNA. That demonstrates the enormous degree of complexity that is represented in the collection of genes from a higher organism."

The development that cut through this complexity was the discovery of enzymes that could slice DNA in specific locations. With these enzymes, researchers became able to isolate specific segments of DNA and reinsert them into other segments of DNA. "By the application of this technology, we can reduce this enormous complexity to relative simplicity," says Leder. "We can reach in through these thousands and thousands of genes and pick out the ones that we are interested in."

The basic technique of recombining DNA is now fairly well established, although its application in the laboratory still entails considerable technical difficulties. First researchers isolate one or more segments of DNA from a living organism, or they chemically synthesize small strands of DNA from its basic constituents. This DNA is usually then spliced into the DNA of a vector, which is most often DNA from a virus; small, independently replicating loops of DNA known as plasmids, which occur in most bacteria and yeast; or genetic combinations of the two, known as cosmids. This genetically engineered vector is introduced into a host cell, which can then reproduce the DNA many

THE GENETIC ENGINEERING OF BACTERIA

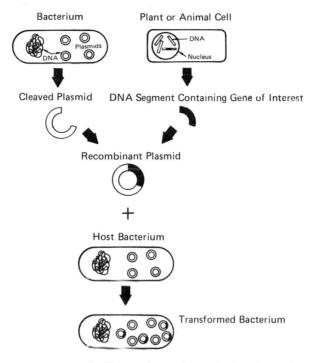

One common way to genetically engineer bacteria involves the use of small, independently replicating loops of DNA known as plasmids. Certain enzymes can cleave these plasmids at specific sequences in their genetic codes. DNA from other organisms that has been treated with the same enzymes can then be spliced into the plasmids with enzymes that join the cut ends of DNA. These recombinant plasmids are reinserted into bacteria, where they can reproduce themselves many times over. At the same time, the bacteria can divide, creating millions of copies of the introduced DNA. This DNA can then be studied through analytical techniques, or, if a gene within the introduced DNA can be made to produce the same protein it did in its original location, the genetically engineered bacteria can be used as microbial factories to make large quantities of the protein.

times over, either for further study or for the production of bioengineered products.

A central concern of researchers has been what causes a gene to produce, or express, the protein it encodes. Unlike the genetic code, the signals that regulate the expression of a protein, which are also encoded in DNA, vary from species to species. Thus, if a human gene is to function in a bacterium, the regulatory signals appropriate to the bacterium must somehow be associated with that gene. This is an important consideration in the industrial application of genetic engineering, since many of the prospective products of biotechnology are proteins. Researchers are working on ways to enhance the expression of a protein by increasing the number of copies of a gene in a cell or by controlling the regulation of the gene.

Research is also being conducted on ways to alter the DNA within genes to yield proteins with improved properties. For instance, enzymes might be modified so that they will catalyze reactions over a broader range of temperatures or chemical conditions. With the use of computers it may even be possible to design enzymes that catalyze entirely new kinds of reactions. This so-called protein engineering could also lead to such advances as storage proteins in plants with more nutritious combinations of amino acids, or new kinds of fibers, plastics, and other materials.

The Molecular Products of Recombinant DNA Technology

The commercial sector that has been most affected by biotechnology is the pharmaceuticals industry. Most of the pharmaceutical products that can be made by genetic engineering are high-value-added substances, which offers an incentive for the large amounts of research and development required to bring them to market. Also, the pharmaceuticals industry has considerable experience with biological processing, since about a fifth of its sales are of products manufactured wholly or in part by microorganisms.

The first therapeutic agent produced through recombinant DNA techniques to be approved by the Food and Drug Administration and to be marketed was human insulin—a protein hormone 51 amino acids long. Human insulin differs from porcine and bovine insulin, which most diabetics use, by only one and three amino acids, respectively. But researchers are hopeful that the use of human insulin will eliminate some of the problems associated with regular injections of animal insulin, including occasional allergic reactions and long-term medical complications; these advantages have not yet been demonstrated in clinical tests. The use of insulin also demonstrates a problem common to all proteins: they must generally be injected under the skin, because proteins taken orally are broken down in the digestive system before they can reach the bloodstream. Work on new kinds of drug delivery systems is therefore an important adjunct to developments in biotechnology.

Another hormone that has been produced through genetic engineering is human growth hormone, a 191-amino-acid protein normally secreted by the pituitary gland. An underproduction of human growth hormone can cause certain kinds of dwarfism, which regular injections of the hormone can at least partially prevent. Clinical trials are being considered to determine if the hormone has additional therapeutic uses.

A number of other protein hormones are potential candidates for production by biotechnology, such as several calcium regulators that may be useful in treating bone disorders, various reproductive hormones, and a number of growth factors that stimulate the development of specific kinds of cells.

A group of hormonelike molecules that have received considerable attention from the biotechnology industry are the interferons, a class of lymphokines. The interferons are glycoproteins—proteins bound to sugar groups—that regulate the body's immune response. They have shown some promise of preventing viral infections, and some evidence suggests that they may be effective in checking certain kinds of infections and cancers. However, it has not been possible to conduct clinical tests to substantiate these claims until recently, when large amounts of interferons became available through genetic engineering. Even if tests do not demonstrate an effective preventive or therapeutic role for interferon, the production of lymphokines through biotechnology promises to reveal much about the functioning of the immune system, which may in turn point the way toward other therapeutic agents.

Proteins fractionated from the human blood represent a substantial market for the pharmaceuticals industry, and several of these have been targeted by the biotechnology industry. Human serum albumin, a protein of 585 amino acids, is used during surgery and to treat shock, burns, and other physical trauma. Antihemophilic factors, specifically factors VIII and IX, are used by the approximately 14,000 hemophiliacs in the United States to control bleeding. And tissue plasminogen activators have demonstrated a remarkable ability to dissolve blood clots in the moments after a heart attack.

The production of large quantities of previously scarce proteins may, in the long run, have a much greater impact than their direct therapeutic uses would indicate. Once these proteins are abundantly available, researchers can study their structure and determine how they function in the body. This information can in turn lead to the design and production of new drugs—whether proteins or nonproteins—to combat disease. This is one example of how genetic engineering can be used as a research tool to probe life's basic processes, with striking implications for health care.

Another important role for proteins in health care may be as vaccines. Today many vaccines consist of the disease-causing organisms in a weakened, or attenuated, state. Once these invaders are injected into the body, the immune system generates antibodies against them and primes itself for future infections of the organisms. However, because the vaccines contain the entire genetic material of the virulent organism, there is a slim chance of contracting the disease from the vaccine. Also, vaccines do not always immunize a person against all strains of a pathogen, and they often need to be refrigerated, making them difficult to use in some parts of the world.

The use of subunit vaccines may solve many of these problems, while also offering the possibility of vaccinating people against a much broader range of diseases. Subunit vaccines consist of just part of a virulent organism, such as part or all of a surface protein. If less than about 50 amino acids long, subunit protein vaccines can be chemically synthesized from their constituent amino acids; proteins of these lengths and longer can also be biologically synthesized by genetically engineered microorganisms. If properly delivered to the body, subunit vaccines can generate an immune response powerful enough to protect against infections by the organism itself. In addition, they can be purer, more stable, and less dangerous than existing vaccines.

A number of viral diseases, including influenza types A and B, herpes, polio, hepatitis, and acquired immune deficiency syndrome, are currently being investigated to determine whether they can be prevented by subunit vaccines. However, it may not be possible to make vaccines for all these diseases. The surfaces of some viruses frequently change, so that previously effective vaccines lose their punch. Ways must also be found to strengthen the immune response that subunit vaccines generate. But even partial successes could have a dramatic effect on health. There are 80,000 to 100,000 cases of hepatitis B in the United States each year, causing about 1,000 deaths, and the incidence of the disease is much greater in other parts of the world. AIDS has already killed thousands of people in the United States, and the incidence of the disease is increasing rapidly.

The development of vaccines against bacterial and parasitic pathogens is more difficult than the development of vaccines for viruses because of the complex and varying surfaces and involved lifecycles of these organisms. For instance, malaria, the most common infectious disease in the world, is caused by a parasite that exists in three different forms in the human body, complicating its prevention by a vaccine. But it may be possible to reproduce surface proteins that occur on some parasites and bacteria and use them as vaccines. It may also be possible to genetically engineer nonpathogenic forms of these organisms that would generate an immune response when injected into the body.

Another use of biotechnology in the pharmaceuticals industry is in the manufacture of metabolites and other nonprotein substances whose reactions are catalyzed by enzymes. For instance, genetic engineering could lead to the production of new antibiotics or make the production of known antibiotics more efficient. Similarly, enzymatic processes could be developed to conduct an increasing number of the chemical steps involved in the synthesis of a wide variety of useful drugs.

Enzymes are important tools in commercial sectors other than the pharmaceuticals industry. Proteases, amylases, and glucose isomerase, for example, are used in food processing and in the manufacture of textiles, detergents, and leather. Two amylases and glucose isomerase are used to convert starch to high-fructose corn syrup, a substance that has increasingly replaced table sugar in processed foods since the late 1960s. Biotechnology may be used both to improve the properties of these and other enzymes and to increase their production from the microorganisms that make them.

The constituents of proteins—amino acids—are another potential product of biotechnology. Amino acids are used as additives in animal feed and human food and for enteral and intravenous feeding. Glutamic acid, whose sodium salt is monosodium glutamate (MSG), and most commercial lysine are now manufactured by strains of *Corynebacterium*. Researchers are applying genetic engineering to this bacterium in an attempt to increase its productivity or give it other desirable characteristics. Tryptophan and phenylalanine are two more amino acids whose economics of production may favor biotechnology.

A number of other metabolites and related high-value compounds may eventually move away from synthetic processing and toward biological processing. Fatty acids and alcohols, vitamins, high-grade oils, flavors and fragrances, adhesives, water-soluble gums, dyes, cosmetics, and many other substances are candidates for production by genetic engineering.

As experience with biotechnology accumulates and production methods get cheaper, less expensive chemicals may also be made through biological methods. Today almost all the commodity chemicals used in industry, serving as precursors for products ranging from solvents to plastics, are synthesized from petroleum and natural gas feedstocks. Essentially all these chemicals could be made from biomass, such as starch or cellulose, and most of them could be produced with microorganisms. Biological processes cannot yet compete with synthetic methods for these chemicals, but this may change as biotechnology advances and the cost of fossil fuels rises.

Fermentation Technologies

All the products mentioned in the previous section are made by what is known as a fermentation process. In this process, living cells or enzymes are combined with nutrients and/or the substance to be chemically transformed in some sort of reactor vessel. The nutrients may consist of sugars, starches, vegetable oil, or even petroleum fractions, and cells may also need additional nitrogen, phosphorus, oxygen, vitamins, metals, or other compounds to grow. Once the desired conversion has taken place, the products of the reaction are removed from the vessel, and the specific compound desired is separated from wastes and by-products and purified for use.

The kind of fermentation technology that now dominates the pharmaceuticals and specialty chemicals industries is batch processing, in which the necessary ingredients are combined in a bioreactor, the conversion takes place, the vessel is emptied, and the entire process begins again. However, continuous processing, in which nutrients and feedstocks enter a bioreactor and spent medium and products leave it on a continuous basis, offers significant advantages for many fermentation products. Continuous processing can have higher productivities and lower costs, because the cells or enzymes are continuously reused and the product is often easier to separate from the outflow. It generally requires, however, that the cells or enzymes be immobilized within the reactor so that they are not swept out with the product. A number of methods have been devised to do this, including bonding the cells or enzymes to a solid support, trapping them in a polymer matrix, or encapsulating them within semipermeable membranous spheres.

The scaling-up to industrial levels of fermentation processes using genetically engineered cells involves a number of difficulties. Maintaining a homogenous mixture of nutrients and dissipating the large quantities of heat generated during fermentation are much more difficult in a full-scale industrial bioreactor than in a small benchtop flask. Also, the bioreactor and incoming nutrients usually have to be thoroughly sterilized, since contaminants can destroy the cells or enzymes or introduce impurities into the final product. This requirement complicates the monitoring of the ongoing reaction, since sensors of many useful measures are disabled by steam sterilization (by the same token, biotechnology may be used to produce sensors that overcome this limitation). Genetically engineered cells can also mutate

Once genetically engineered microorganisms have been grown in a fermentation process, their components are separated using the centrifuges shown here. Further separation is then required to isolate the one protein that is desired from the thousands of other proteins produced by the microorganism. The isolated protein must be rigorously purified to eliminate contaminants from the final product. In many cases, this separation and purification process is more expensive than the original fermentation.

or revert to an earlier genetic state during fermentation, making the products of the fermentation useless.

The separation and purification of products from dilute aqueous solutions presents another set of problems. In many pharmaceutical applications, this phase of production costs more than the fermentation itself. In addition to such standard techniques as distillation, drying, and precipitation, bioprocess engineers are experimenting with the use of ultrafiltration, high-performance liquid chromatography, electrophoresis, and antibody technology to recover products.

There is a pressing need in biotechnology for microorganisms better suited to fermentation technologies. For instance, the bacterium *Escherichia coli*, which has been widely used in genetic engineering, manufactures its products intracellularly and also produces highly toxic substances called endotoxins that must be rigorously eliminated from the final product. Researchers are investigating other kinds of bacteria and higher microorganisms like yeast that can be induced to secrete their products into the surrounding medium and that do not produce toxic compounds.

Ways must also be developed to grow other kinds of cells, including plant, animal, and human cells, in cultures for industrial purposes. These cells will ultimately be the most useful producers of many valuable substances in biotechnology. However, their nutritional requirements are poorly defined, and they are much more fragile and complex, and hence more difficult to grow, than are one-celled microorganisms like bacteria and yeast.

Microorganisms for Use in the Environment

In addition to their uses in fermentation processes, genetically engineered microorganisms will find direct application in the environment. The agricultural uses of genetically engineered organisms are discussed in Chapter 3. But that leaves a variety of other industrial processes to which biotechnology could contribute.

Liquid and solid wastes are broken down in waste treatment plants largely through the action of microbes. Biotechnology could produce enzymes or other substances that hasten or further this process. For example, biologically derived flocculants would be very useful for separating and thickening solids during treatment. Cellulases, proteases, amylases, and polysaccharide hydrolases could help release the water retained in sludge before it is disposed of. It may even be possible to genetically engineer properties into microorganisms that would enhance their ability to break down certain waste substances— not only sludge but slime, grease, and scum as well.

Genetically engineered microorganisms or their products may also be able to remove heavy metals or organic pollutants, including suspected carcinogens, from drinking water and industrial wastewater. Proteins known as metallothioneins can bind various kinds of heavy metals, and other proteins can polymerize aromatic compounds so that they can be removed by flocculation. Microbiologists have either found or produced through conventional genetic techniques organisms that can break down a variety of toxic substances, including 2,4-D and 2,4,5-T. Once the genes controlling these processes are isolated and characterized in an organism, they could be transferred to other organisms via recombinant DNA.

The ability of enzymes to recognize and bind metals is important in another possible environmental application of biotechnology: microbial mining. Microorganisms are already used for leaching low-grade ores and concentrating metals; in fact, more than 10 percent of the copper produced in the United States is leached from ores by microbes. Genetic engineering could improve these organisms in any number of ways: by increasing their tolerance to saline or acidic conditions, by decreasing their toxicity to certain metals, or by increasing their ability to withstand high temperatures in underground mines.

Another use for microorganisms or their products might be to enhance the extraction of oil from wells. Only about half the world's supply of subterranean oil reserves can be recovered using conventional techniques. Biologically derived surfactants and viscosity decreasers could be injected into wells to enable some of this additional oil to be pumped out. Furthermore, if organisms were found or genetically engineered that could live under the harsh conditions of oil wells and give off the proper products, they could be directly introduced into wells to repressurize or condition the oil for removal. However, as with many of the other environmental applications of biotechnology, considerable additional research is necessary before this will be possible.

Monoclonal Antibodies

Recombinant DNA is just one of the techniques that have led to the development of biotechnology over the past decade. A panoply of other procedures, from protein sequencing to tissue culturing, have also contributed to the growth of the field.

One of the most prominent of these procedures is cell fusion. In this process, the constituents of two different cells are combined to form a single hybrid cell. Cell fusion has given rise to a variety of exotic organisms, such as the hybrid plants mentioned in Chapter 3. But the most important outgrowth of cell fusion, accounting at this point for more commercial products than recombinant DNA has generated, is the production of monoclonal antibodies.

Antibodies are complex proteins that are produced and secreted by B lymphocytes, a type of white blood cell that forms an important component of the body's immune system. Antibodies have the ability to recognize and attach themselves to foreign substances in the body—known collectively as antigens—setting in motion a process that will eliminate the antigen from the body. Each lymphocyte produces only a single kind of antibody, but there are a virtually unlimited number of different lymphocytes, and each proliferates rapidly when it detects its corresponding antigen. In this way the immune system offers protec-

CELL FUSION

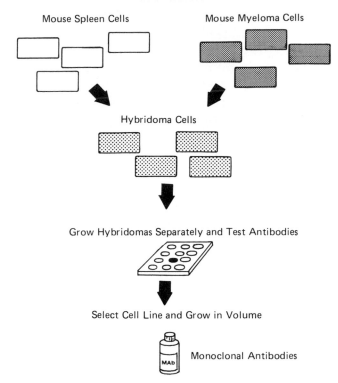

Mouse Spleen Cells   Mouse Myeloma Cells

Hybridoma Cells

Grow Hybridomas Separately and Test Antibodies

Select Cell Line and Grow in Volume

Monoclonal Antibodies

To produce monoclonal antibodies, antibody-producing spleen cells from a mouse that has been immunized against an antigen are mixed with mouse myeloma cells. Under the proper conditions, pairs of the cells fuse to form antibody-producing hybrid-myeloma ("hybridoma") cells, which can live indefinitely in culture. Individual hybridomas are grown in separate wells, and the antibodies they produce are tested against the antigen. When an effective cell line is identified, it is grown either in culture or in the body cavities of mice to produce large quantities of chemically identical, monoclonal antibodies.

tion against a wide range of infectious agents and other foreign substances.

The traditional means of producing antibodies for research and medical purposes has been to inject an animal with an antigen and collect the antibodies that result. This method has several drawbacks, however. The injection of an antigen generates many different antibodies that react with the antigen; the supply of antibodies produced in this way is limited; and an injected antigen usually contains other materials, leading to the production of antibodies against a variety of antigens.

In 1975 Cesar Milstein and Georges Köhler of the British Medical

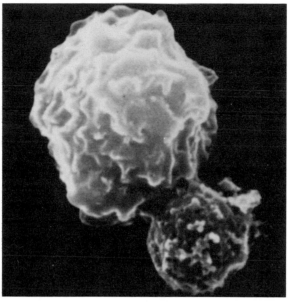

A mouse spleen cell and tumor cell fuse to form a hybridoma. As the hybridoma divides, it gives rise to a "clone" of identical cells, giving the name "monoclonal" to the antibodies those cells produce.

Research Council Laboratory of Molecular Biology in Cambridge happened on a way around these problems. They fused myeloma tumor cells from a mouse, which have the capacity to grow indefinitely in culture, with cells derived from mouse B lymphocytes, which have a limited lifetime. The resulting hybrid-myeloma or hybridoma cells combined just the right qualities of each parent. They prospered in cell culture and at the same time produced virtually unlimited quantities of chemically identical or monoclonal antibodies (so named because they are produced by the cloned copies of a single hybridoma).

Monoclonal antibodies have already begun to find many valuable applications in research, according to William Paul of the National Institute of Allergy and Infectious Diseases at the National Institutes of Health. Because of their great specificity, they are ideal tools for separating and purifying proteins and other cellular components, tasks that were often difficult with conventional techniques. "In certain circumstances, one might even consider using monoclonal antibodies for large-scale [industrial] purification," says Paul, and in fact it is already being used for such purposes. However, this is only possible "where it is economically feasible to do so, which probably limits its use to molecules that are active in exquisitely low concentrations and that would be difficult to purify by more conventional means."

Monoclonal antibodies also have "enormous potential" in medicine, according to Paul. "It is in principle possible to develop a measurement technique for essentially any molecule that is immunogenic or could be made immunogenic through chemical manipulation," he points out. This has already led to in vitro diagnostic tests for detecting and monitoring pregnancy, venereal diseases like chlamydia and herpes, viral infections leading to hepatitis and AIDS, bacterial infections causing meningitis, and some forms of cancer.

Research is also being conducted on the use of monoclonal antibodies in vivo, although the more rigorous safety testing demanded of such products has so far limited their use. Most obviously, monoclonal antibodies can be used to confer short-term passive immunity, as opposed to the long-term active immunity conveyed by lymphocytes. Researchers are also investigating the use in the body of monoclonal antibodies tagged with radioisotopes or other chemicals. By binding to blood clots or tumors in the body, such constructs could reveal the location of these pathologies to scanning devices. Monoclonal antibodies injected into the body may be able to halt the spread of certain tumors, and research is being conducted into the possibility of attaching toxic agents to the antibodies that would be delivered directly to a tumor. "The great difficulty is the preparation of antibodies that truly distinguish tumor cells from normal cells," says Paul. "At the current time the numbers of situations in which really good results may be obtained are very limited."

Hybridoma technology faces other limitations that have hampered its full effectiveness. For one thing, it remains hard to create monoclonal antibodies against antigens that generate only a weak immune response. Also, it has proved very difficult to create hybridomas from human cells. "The monoclonal antibodies that have been produced thus far of great value have been derived ultimately from either mice or rats," says Paul. "Successes with cells from more distantly related animals have been rare indeed, and those results are very disappointing."

In the future it may become possible, using recombinant DNA techniques, to genetically engineer cells to produce unlimited quantities of specific antibodies. But for the immediate future, researchers are working on forming hybridoma cell lines from human cells rather than from mouse cells. Such cell lines would have a number of advantages. The body would be less likely to generate an immune response against human antibodies than against mouse antibodies. Also, human monoclonal antibodies might have even greater specificities than those now available.

Pairs of human tumor cells and human lymphocytes have been induced to fuse in laboratories, but the resulting hybridomas are difficult to grow in culture and tend to be genetically unstable. "Although some successes have been achieved, and it would be wrong to discount those successes, this technology has not yet reached a state in which one can reliably produce human monoclonal antibodies," says Paul. "I have very little doubt that, with the very large number of individuals who have great interest in these areas, progress will be made. But we should not overemphasize how far we have come along what is an exciting but still very difficult pathway."

Additional Readings

Cesar Milstein. 1980. "Monoclonal Antibodies." *Scientific American* 243(October):66-74.

Office of Technology Assessment. 1981. *Impacts of Applied Genetics: Micro-Organisms, Plants, and Animals*. Washington, D.C.: U.S. Government Printing Office.

Office of Technology Assessment. 1984. *Commercial Biotechnology: An International Analysis*. Washington, D.C.: U.S. Government Printing Office.

Scientific American. 1981. *Industrial Microbiology and the Advent of Genetic Engineering*. San Francisco: W. H. Freeman.

James D. Watson, John Tooze, and David T. Kurtz. 1983. *Recombinant DNA: A Short Course*. New York: W. H. Freeman.

3

Biotechnology in Agriculture

WITH THE EXCEPTION OF THE pharmaceuticals industry, agriculture is the commercial sector that has drawn the most attention from biotechnology. Many of the applications of genetic engineering in agriculture will probably take longer to emerge than in other areas, given the complexity of some of the problems that must first be solved. But the potential returns are unparalleled. Furthermore, several developments of the past few years have indicated that progress may be more rapid than was previously thought possible.

Many of the products described in Chapter 2 for use in human health care have agricultural analogs. For instance, hormones, steroids, and antibiotics, which are all potential products of biotechnology, have long been used in agriculture. Biotechnology will also yield drugs, feed additives, and growth enhancers that have never before been available in commercial quantities. A particularly intriguing example is growth hormone, which may result in faster growing, larger, and leaner animals and which has increased milk production in dairy cattle by up to 40 percent.

Products created through genetic engineering will also be used to diagnose and treat disease, which each year reduces the productivity of

---

This chapter includes material from the presentations by Ernest G. Jaworski, Rudolf Jaenisch, and Philip Leder at the symposium.

livestock and poultry in the United States by 20 percent. A genetically engineered vaccine for colibacillosis or scours, a diarrheal disease that kills millions of newborn calves and piglets each year, is already available. A subunit vaccine has been genetically engineered for foot-and-mouth disease in cattle, which remains a serious problem throughout South America, Africa, and the Far East. Other vaccines are being developed for rabies, swine and canine parvovirus, fowl plague, bovine papilloma virus, and many other diseases.

Certain products of biotechnology will be used to detect and monitor the progress of disease, so that treatment can begin before economic losses occur. Researchers are developing monoclonal antibodies to diagnose bluetongue (a viral disease in sheep transmitted by gnats), equine infectious anemia, bovine leukosis virus, and a number of viral diseases that strike dogs and cats. Monoclonal antibodies can also fend off disease by conferring passive immunity to an infectious agent. Furthermore, as is the case throughout genetic engineering and biotechnology, researchers use such tools as monoclonal antibodies to learn more about the origins and mechanisms of diseases, which can in turn point toward more effective therapies.

All the products mentioned above are made through fermentation processes or cell culture techniques. But biotechnology has another, fundamentally different, capability in agriculture. It can be used to genetically alter agriculturally important animals, plants, and microbes, producing crops and livestock with characteristics that cannot be achieved through traditional breeding programs. For instance, microorganisms might be genetically engineered that provide nitrogen to important crops, greatly reducing the need for fertilizer. Plants might be produced that grow faster or in more places or that have larger and more nutritious yields. Animals might be able to secrete elevated levels of their own growth hormone so that they grow faster and larger. These are the truly revolutionary agricultural applications of biotechnology, and they are the subject of this chapter.

Genetically Engineered Microorganisms in Agriculture

Microorganisms in the environment affect the growth of plants and animals in a variety of ways, many of which are still poorly understood. As research progresses, it should be possible to genetically engineer these microorganisms to yield hardier and more productive crops and livestock. Given the unresolved difficulties involved in altering the genetic material of plants and animals, this may be the first direct application of genetic engineering in agriculture.

The best known and most intensively studied relationship between microorganisms and plants involves the essential nutrient nitrogen. Plants cannot directly absorb and use the nitrogen gas that constitutes more than 75 percent of the atmosphere. It must first be fixed or converted into other nitrogen-containing compounds, either in industrial facilities that produce fertilizer or in certain bacteria and blue-green algae that live in the soil. The most agriculturally important nitrogen-fixing bacteria belong to the genus *Rhizobium*. These bacteria infect the roots of members of the legume family, including beans, peas, soybeans, peanuts, alfalfa, and clover, providing the plants with nitrogen and symbiotically receiving nourishment from the plants. This buildup of nitrogen-containing substances in turn increases the fertility of the soil for nonleguminous crops, an observation made as far back as Roman times.

Researchers are trying to genetically engineer *Rhizobium* bacteria so that they will fix nitrogen more efficiently or infect other crops in addition to legumes. An important consideration in this work is the competitiveness of the genetically engineered bacteria. More productive *Rhizobium* must be able to survive and to displace indigenous *Rhizobium* if they are to have an effect.

Other microorganisms also fix nitrogen in the environment, and researchers are examining these to see if they could be adapted to supply nitrogen to crops. The challenges involved in this work are to engineer the organisms so that they will live in association with the desired crops and fix excess nitrogen beyond their own metabolic needs.

Alternatively, researchers are investigating the possibility of transferring the ability to fix nitrogen to microorganisms that already live in association with a given crop. The 17-gene complex that enables the bacterium *Klebsiella pneumoniae* to fix nitrogen was isolated, reproduced, and introduced into *Escherichia coli*, which then became nitrogen-fixing. But when the same genes were inserted into yeast cells, no nitrogen was fixed, indicating the difficulties likely to be encountered in trying to transfer this capacity to higher organisms.

Other microorganisms affect plant growth in different ways. Some protect plants from bacterial or fungal infections or secrete compounds that regulate a plant's development. Others protect plants from such environmental conditions as acidity, salinity, and high concentrations of toxic metals. Some microorganisms are able to degrade toxic substances used as pesticides, like 2,4-D. Others can kill weeds or other plants that compete with a crop for nutrients. As these and additional relationships between plants and microorganisms become better understood, genetic engineering will turn to the production of altered microorganisms that enhance the vigor and growth of crops.

U.S. Department of Agriculture

Nitrogen-fixing *Rhizobium* bacteria form nodules on the roots of plants they infect, supplying the plant with nitrogen in return for nourishment from the plant. Genetic engineers are trying to alter *Rhizobium* bacteria so that they will infect plants other than legumes. Alternatively, researchers are seeking to transfer the ability to fix nitrogen into other microorganisms that live in association with crops.

Genetic engineering will also be used to combat those microorganisms, such as certain bacteria and fungi, that harm crops. A particularly interesting example involves the bacterium *Pseudomonas syringae*. A protein on the surface of this widespread bacterium initiates the formation of ice when temperatures drop below freezing. If this protein were eliminated through recombinant DNA or conventional mutational techniques, temperatures could drop several more degrees before frost damage began to occur.

Microorganisms amenable to genetic engineering also play critical roles in animal agriculture. For instance, some microbes are lethal to the insects that transport diseases into animals. An example is the bacterium *Bacillus thuringiensis,* which produces a toxin that is deadly to mosquitoes and black flies. Other microorganisms perform their functions within animals. For example, ruminants can consume forage because it is fermented by microbes in their digestive tracts. It is even possible that the genetic engineering of these microorganisms could give animals the ability to digest foodstuffs that are now useless to them.

The Genetic Engineering of Plants

A more direct way to enhance the productivity of agriculturally important plants and animals is to alter the DNA that dictates their characteristics. At the most basic level, this is what plant and animal breeders have been doing since the dawn of agriculture. In recent decades, plant and animal breeders have developed sophisticated techniques to transfer traits among organisms that can interbreed. They have also developed a host of supporting technologies, such as cell and tissue culture, embryo transfer, and artificial insemination, that facilitate these basic genetic manipulations. In this sense, genetic engineering will be building on a base of experience and expertise that has accumulated over centuries. But at the same time it will offer capabilities that have never before been available.

According to Monsanto's Ernest Jaworski, three things are needed for the genetic engineering of plants: a host cell or tissue, a vector to transfer DNA into the host, and the segments of DNA that are to be transferred.

Protoplasts have been a popular choice for hosts in the genetic engineering of plants. Protoplasts are cells taken from the leaves, stems, or roots of a plant that have been exposed to enzymes that dissolve the cells' tough outer walls. The "nakedness" of these cells makes it much easier to introduce DNA into them.

The use of protoplasts as hosts is critically dependent on their ability to give rise to whole plants, a characteristic known as totipotency. Through exposure to the proper nutrients and plant hormones, protoplasts can be induced to regenerate cell walls and undergo cell division to form an undifferentiated mass of callus tissue. In some cases, this callus tissue can then be induced to differentiate into shoots, roots, or entire plants. However, it is not yet possible to regenerate whole plants from callus tissue for most of the agriculturally important food crops, and the factors controlling this process are still poorly understood.

Once a protoplast host has been prepared, foreign DNA can be inserted into the cell in several different ways. Two protoplasts can be made to fuse, producing a hybrid cell that in some cases can be regenerated into a plant with novel characteristics. For instance, potato and tomato protoplasts have been fused to produce a hybrid dubbed the "pomato." Discrete segments of DNA, in the form of chromosomes, whole nuclei, or cell organelles (some of which contain their own nonnucleic DNA), can also be inserted into a cell mechanically. But the most powerful and versatile way of introducing DNA into a plant cell hinges on the properties of an unusual plant pathogen.

"Nature was very kind to the plant molecular biologists," explains Jaworski. "It supplied us with a natural, soilborne organism called *Agrobacterium tumefaciens*. This soilborne organism invades plant tissues through wound sites and introduces genetic information, by a mechanism unknown as yet, into the chromosome of the plant cell. . . . This is one of the greatest systems for transforming plants that has been invented to date."

Plant researchers have discovered that the agent in *A. tumefaciens* enabling it to perform this transformation is a large plasmid that has the ability to insert part of its DNA at a random location into the DNA of the cell nucleus. Normally, the genes inserted by this plasmid code for plant hormones that cause tumors in plants known as crown galls. But through the use of genetic engineering, researchers have deleted those tumor-inducing genes and have inserted genes of their own choosing. The first gene to be inserted in this way and expressed in a whole plant—in this case a petunia—was a gene conveying resistance to an antibiotic. Even some of the offspring of these plants were resistant to the antibiotic, demonstrating that the new DNA was passed on as a stable genetic entity.

The genetic engineering of plants is clearly still in its infancy, but the early success of genetic engineering in some plants points the way toward a time when it may be possible to introduce desirable traits into many agriculturally important crops. For instance, researchers at Monsanto and elsewhere have been working with the genes that code for the enzyme ribulose-1,5-bisphosphate carboxylase-oxygenase, often referred to simply as Rubisco. Rubisco, which is probably the most abundant protein in the world, is the key catalyst in photosynthesis, the process that allows plants to convert carbon dioxide from the atmosphere into other carbon-containing compounds the plants can use. Rubisco consists of eight large subunits encoded by genes in chloroplasts and eight small subunits encoded by genes in the nucleus. Although no vector systems exist to alter genes in the chloroplasts, researchers are genetically engineering the genes that encode the

THE GENETIC ENGINEERING OF PLANT CELLS

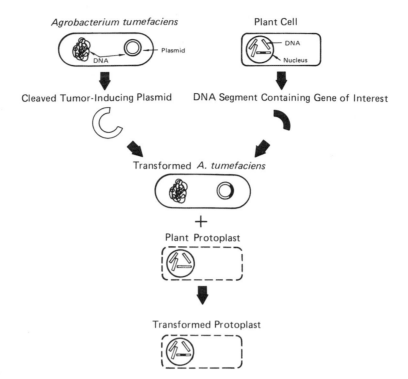

To genetically engineer plant cells, molecular biologists use an unusual soil pathogen, *Agrobacterium tumefaciens*. This bacterium contains a large tumor-inducing plasmid that can insert part of its DNA into the DNA of whatever plant cell the bacterium infects. If researchers replace the tumor-inducing genes of the plasmid with foreign DNA, the bacterium can be used as a vector to introduce novel DNA into plant cells. (Most often these cells are protoplasts, normal plant cells that have been exposed to enzymes that dissolve their cell walls, making it easier to introduce DNA into them.) Once transformed by *A. tumefaciens*, the cells can sometimes be regenerated into whole plants through exposure to the proper combinations of plant hormones and nutrients.

small subunits. By increasing the efficiency with which Rubisco fixes atmospheric carbon dioxide, researchers hope to produce plants that will grow faster.

In collaboration with researchers at Rockefeller University, plant molecular biologists at Monsanto have introduced into petunias the genes coding for Rubisco's small subunits in the pea. "It actually gets produced, processed and transported properly, and assembled as part of the petunia Rubisco holoenzyme," says Jaworski. "That, I think, is very encouraging." Furthermore, the bioengineered petunia Rubisco demonstrates many of the properties of Rubisco in the pea. "The gene is light-dark regulated: it is turned on by the light, it is not on in the dark. [Also] the gene is only expressed in the appropriate tissue. We did not find this pea small-subunit gene being expressed, for example, in the roots and stems of the plant, but only in the leaves."

Another group of proteins that have been a focus of work by plant molecular biologists are the storage proteins in plant seeds. The seeds of legumes and cereal grains provide humans with an estimated 70 percent of their dietary protein requirements, but some of the most important storage proteins in these seeds are deficient in certain essential amino acids that must be made up in other ways. Researchers have consequently examined the possibility of genetically engineering the genes that code for these proteins to alter their amino acid composition.

Such efforts quickly run up against a number of difficulties, according to Jaworski. "The storage proteins in crops such as corn and soybeans are very complex, multigene families. There are a number of pieces of information we don't have about exactly what happens when you genetically engineer a protein. Let's say we modify it with a single amino acid change. We don't know how that might affect the secondary and tertiary structure of the protein, which has to do with how it is going to be folded and deposited when it is being formed as a storage protein."

Jaworski believes that a technically more feasible goal is the modification of leaf proteins rather than storage proteins, which are also commercially valuable as feed for livestock. It might be possible to alter these proteins to be more nutritious, or their concentration in leaves might be amplified through genetic engineering. But this, too, encounters certain difficulties. "We don't know what happens when we elevate natural proteins beyond a certain level," says Jaworski. "We know that in bacteria this can be lethal."

The number of genes that are involved in determining a given characteristic is crucial to whether that characteristic is amenable to genetic engineering. For instance, it would be desirable in many cases to give crops the ability to fix their own nitrogen or to photosynthesize more efficiently. Crops might also be genetically engineered to produce higher levels of plant growth hormones or to have a greater ratio of harvestable to nonharvestable matter. Resistance to such environmental factors as disease, insects, competing plants, flooding, drought, salinity, toxic metals, pesticides, heat, and cold are all potential goals of biotechnology. Unfortunately, many of these attributes are probably the result of the interaction of many genes, making them difficult to decipher and transfer from one plant to another.

But at least some of these traits *are* thought to be controlled by one or a handful of genes. For instance, researchers at Monsanto are trying to isolate the gene that codes for the enzyme EPSP synthase, which enables plants to resist an herbicide known as glyphosate that is sold by Monsanto under the trade name Roundup. By growing plant cells in the presence of increasing levels of the herbicide, the researchers were able to isolate a strain with a greatly amplified expression of the gene. They have now reproduced this gene and are trying to introduce it into new plants. Similarly, other plant researchers are trying to isolate and transfer genes that enable plants to make their own insecticides, resist infection by pathogens, or stand up better to a variety of environmental stresses. If successful, says Jaworski, "we can certainly change the geography of some of the cropping practices that limit us today to only specific areas of the world."

Before these advances become a reality, several technical problems must be overcome. First, regeneration of whole plants from single-celled protoplasts has so far been accomplished only in a limited number of dicotyledons (flowering plants with two seed leaves), including tomatoes, tobacco, potatoes, and petunias. Regeneration has generally not been possible with monocotyledons, the group of flowering plants that includes the important cereal grains. Work is under way to develop regeneration systems for these crops, but the continued lack of such systems will severely limit the range of plants that can be genetically engineered.

Similarly, *A. tumefaciens* will only infect dicots, although plasmids similar to the one it carries might be induced to transform monocots. Consequently, researchers are intensively searching for other kinds of vectors that can introduce foreign DNA into plants. Pieces of DNA that can move about within the genome, known as transposons, are one possibility. Investigators are also looking at geminiviruses, which are single-stranded DNA viruses that may be able to transform some plant cells.

But the most fundamental problem in applying genetic engineering to agriculture, according to Jaworski, is a lack of basic biochemical knowledge about plants. "We need to spend a lot more time—and this is where I think we will see a great deal of activity in the next five to ten years—on identifying agronomically important traits and the genes that regulate those traits," he says. "If we cannot do this, we are not going to be very successful in really making the agronomic improvements that we desire to make." Even with such well-studied functions as photosynthesis, much more work needs to be done to understand the biochemical pathways of regulation at the genetic level. "We need a great deal more information about the signals that regulate tissue specificity, developmental specificity, temporal specificity, and so on," says Jaworski. "We just don't have enough knowledge yet to understand how to regulate at will, and in a controlled fashion, the expression of a gene."

The research needed to acquire this knowledge requires both greater cooperation between plant molecular biologists and traditional plant breeders and a commitment by the federal government to fund this kind of interdisciplinary effort, according to Jaworski. "There is a lot of basic research that has to be done in parallel with the applied research if we are going to be successful in moving the technology from the laboratory into the field."

The Genetic Engineering of Animals

Unlike plants, an animal cannot be regenerated asexually from cells plucked at random from certain parts of its body. Only one kind of cell—the zygote formed by the fusion of a sperm cell and an egg—has the capacity to develop into a fully formed animal. Therefore, to introduce a foreign gene into *all* the cells of an animal, including the germline cells that will pass on an animal's genetic heritage to its offspring, the foreign DNA must be inserted into the sperm, the egg, or the zygote. If a multicell embryo is exposed to foreign DNA, the resulting animal will be a mosaic—some of its cells will carry the introduced genes and some will not. If foreign DNA is inserted into cells of the organism even later—say, after birth—a correspondingly smaller number of cells will be altered.

There are several ways of introducing specific genes into the chromosomes of an animal's cells, according to Rudolf Jaenisch of the Whitehead Institute for Biomedical Research and the Massachusetts Institute of Technology. One of the most widely used is to insert copies of DNA directly into cells using a micropipette. This seems to work best when done to zygotes after the egg has been fertilized but before the genetic material of the egg and sperm have joined. "The success of deriving transgenic mice in this manner is variable," says Jaenisch. "In a good laboratory between 10 and 30 percent of the animals born will carry the foreign sequences in the germ line."

Another way to transform animal cells with foreign DNA is by using retroviruses as vectors. Retroviruses are infectious agents that cause a wide variety of diseases in humans and animals, including some forms of leukemia in nonhuman species. They have the ability to insert a single strand of DNA, derived from their own genetic material, into the DNA of the cells they infect. By genetically engineering certain kinds of retroviruses, researchers can replace their disease-causing genes with genes coding for other proteins. As with *A. tumefaciens* in plants, retroviruses can then incorporate these genes into the DNA of their hosts.

In 1981 a rabbit globin gene became the first bioengineered gene to be inserted into an animal embryo—in this case a mouse—and reproduced in all the cells of the mature animal. Since then a number of other genes, including oncogenes and genes coding for metallothionein, elastase, and immunoglobulin, have been inserted, expressed, and passed on to offspring in laboratory animals. A landmark experiment was the introduction into mice of a gene for growth hormone fused to the regulating DNA from a metallothionein gene that caused growth hormone to be expressed whenever the mice were exposed to certain heavy metals. The mice transformed by the growth hormone gene grew to more than twice the size of their normal siblings.

The success of these experiments has generated great interest in the possibility of genetically engineering farm animals so that they would be more productive or more resistant to disease. To date, much of this interest has focused on the prospects for growth hormone. Experiments with injected growth hormone have suggested that animals producing elevated levels of their own growth hormone might grow faster, larger, leaner, and with less consumption of feed. Such advances would be particularly welcome in the production of swine, since they are generally sold at an immature age and since consumers would likely favor leaner pork. Injections of growth hormone have also been shown to markedly increase the production of milk in dairy cows.

Jaenisch warns, however, that several questions must be answered before the genetic engineering of farm animals becomes practical. For instance, researchers are still not certain whether elevated levels of growth hormone would have harmful side effects or whether such levels would even produce the increased growth expected. Swine are already bred for maximal growth, and it is not clear whether insertion of a growth hormone gene would further increase their size. Also, the mice transformed by the growth hormone gene showed signs of abnormally proportioned growth, and the female mice genetically engineered in this way were often sterile. "So there are a number of physiological consequences of inserting a gene that we really don't understand yet," Jaenisch says. "I think one has to be cautious about

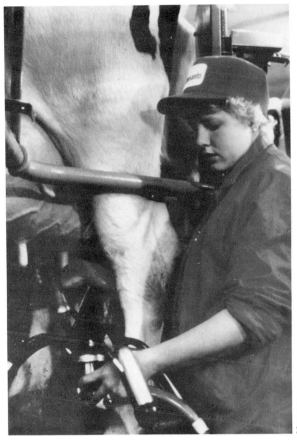

Dairy cows injected with growth hormone increased their production of milk up to 40 percent. Through genetic engineering, researchers hope to create cattle that will produce elevated levels of growth hormone endogenously, resulting in faster growth or increased milk production without the need for injections.

drawing too many conclusions of what the value of this technique will be for general use."

As in the genetic engineering of plants, an even more fundamental problem involves the regulation of the genes inserted into animal cells. The expression of an inserted gene can be influenced both by the regulatory sequences associated with the gene and by where the gene is inserted into the DNA of its host. At present, there is no way to control where a gene is inserted into the chromosome of either an animal or a plant cell. Yet this position of insertion can affect not only the expression of the inserted gene but also the regulation of the host cell's DNA. For instance, inserted DNA can separate two sections of a functioning gene and block its action, causing genetic disease in interbred offspring if the gene is recessive (all of them have been to date). Inserted genes can also turn on even distant genes within the genome, causing a tumor if the activated gene is an oncogene.

Such mutations are valuable in what they tell molecular biologists about the biochemical machinery of genetic regulation. But much more needs to be learned before it will be possible to insert genes in a predictable fashion and control the expression of those genes for desired ends. "The major scientific problem that confronts the geneticist is our inability in higher organisms to predictably, invariably, and inevitably replace and alter genes at will," says Philip Leder of the Harvard University Medical School. "It is not possible for us now to introduce genetic material into the mouse . . . in a way in which the outcome of that experiment is absolutely predictable. It is not yet possible to correctly or predictably alter the amino acid composition of the major corn protein by introducing amino acids that are essential for human and animal nutrition, or to predict if that protein will be expressed in normal, or perhaps larger, amounts. To be able to do that predictably will open a new and important avenue for application and investigation and has to be viewed as one of our major scientific goals."

Additional Readings

Kenneth A. Barton and Winston J. Brill. 1984. "Prospects in Plant Genetic Engineering." Pp. 121-131 in *Biotechnology and Biological Frontiers*, Philip H. Abelson, ed. Washington, D.C.: American Association for the Advancement of Science.

Winston J. Brill. 1981. "Agricultural Microbiology." *Scientific American* 245(September):198-215.

National Research Council, Board on Agriculture. 1984. *Genetic Engineering of Plants: Agricultural Research Opportunities and Policy Concerns.* Washington, D.C.: National Academy Press.

National Research Council, Committee on Biosciences Research in Agriculture. 1985. *New Directions for Biosciences Research in Agriculture.* Washington, D.C.: National Academy Press.

Thomas E. Wagner. 1984. *The Implications of Genetic Engineering in Livestock Production.* Knoxville: University of Tennessee.

4

Human Gene Therapy

IN THE EARLY 1970s an American researcher named Stanfield Rogers infected three German girls who lacked the enzyme arginase with Shope papilloma virus, hoping that the virus would transfer to the girls the gene for the missing enzyme. In 1980 Martin Cline of the University of California at Los Angeles exposed the bone marrow of two patients from Italy and Israel who suffered from beta-thalassemia (a blood disorder resembling sickle cell anemia) to recombinant DNA coding for the blood protein hemoglobin, hoping that the bone marrow would incorporate the new genes and alleviate the patients' disease.

Neither of these first two attempts at human gene therapy had an effect on the patients involved. But they dramatically affected the biomedical community and, especially in Cline's case, the public. They demonstrated that attempts to alter the genetic constitution of human beings were not a distant prospect, sufficiently far off to leave years for exploring their scientific and ethical implications, but a present reality. In the past few years the advent of successful human gene therapy has come even closer. Since 1980 the replacement of defective genes has been accomplished in fruit flies and mice. At least six major research centers in the United States are working to develop the

This chapter includes material from the presentations by W. French Anderson, Leroy B. Walters, Jr., and Albert Gore, Jr., at the symposium.

43

techniques that will permit gene therapy in humans. The first clinical trials of these techniques in human subjects are expected to begin in 1986.

There are many different types of human gene therapy, each with its own set of scientific and ethical questions. The kind of gene therapy now being pursued by researchers is far removed from the kinds of gene therapy that many people, for a variety of reasons, have come to fear. The first sanctioned attempts at gene therapy will involve the insertion of a single gene into a limited subset of a patient's cells to palliate a severe genetic disorder. The gene will not be able to spread beyond those cells, and it will not be passed on to the patient's offspring. Gene therapy of this type closely resembles other types of medical procedures, such as transplants or drug treatment, and a consensus has gradually emerged that it presents no new ethical problems.

Other kinds of human gene therapy can be envisioned, but formidable technical difficulties make it hard to imagine when, if ever, they may become feasible. For instance, researchers have succeeded in changing the genetic constitution of mice so that new genes are passed down from generation to generation. But it is not now possible to do this with humans, and technical and ethical problems inherent in such work are likely to keep it from being attempted for years.

Any endeavor to genetically engineer human beings to enhance certain characteristics (versus repairing an inborn genetic defect) would of course raise difficult questions of ethics and safety. For instance, it is impossible to predict how the introduction of one or a few "enhancing" genes into the body's cells would affect the health of either an individual cell or an entire person.

Such complex human attributes as intelligence, character, and physical appearance are undoubtedly controlled by many genes interacting among themselves and with innumerable environmental influences. It is difficult to conceive of how genetic engineering could ever be used to affect these complex human traits. But that does not mean that the broad moral and social implications of human gene therapy should not be the subjects of continuing reflection. As the experiments of Rogers and Cline demonstrate, events in biomedical research have often leapt ahead of their ethical and philosophical underpinnings.

Somatic Cell Gene Therapy

The kind of gene therapy now being studied by researchers involves the insertion of one or a handful of genes into somatic cells in the body. Somatic cells include all the body's cells except for sperm cells, egg cells, and the cells that give rise to them, which are collectively known as germline cells. Because somatic cell gene therapy does not affect the germ line, the genes conveyed through the procedure will not appear in the recipients' offspring.

The first diseases selected for treatment with somatic cell gene therapy will share several characteristics, according to W. French Anderson of NIH's National Heart, Lung, and Blood Institute. First, they will arise from a defect in a single gene causing the loss of an enzyme with potentially lethal consequences. "Those genetic disorders that are serious but not lethal are unlikely to be the first candidates," says Anderson. Defects in single genes cause more than 200 known human disorders, including muscular dystrophy, sickle cell anemia, cystic fibrosis, and hemophilia, and there are more than 2,000 known genetic diseases. But only a few of the genes responsible for single-gene disorders have so far been isolated and reproduced through genetic engineering so that copies of them can be inserted into cells.

Second, the diseases will be treatable through the genetic manipulation of bone marrow cells, because techniques have been developed to remove these cells from the body, transform them with recombinant DNA, and reintroduce them into the body. Perhaps in the future it will be possible to genetically manipulate skin cells and even tissues and whole organs, but for now bone marrow cells are the only cells conducive to this kind of treatment.

Finally, the genes responsible for the diseases will have a fairly simple kind of regulation. It was originally thought that the various diseases caused by defects of hemoglobin, such as sickle cell anemia and beta-thalassemia, would be the first disorders to be treated with gene therapy. However, the regulation of hemoglobin production has turned out to be unusually complicated, involving several different genes on different chromosomes. Thus, the first genes to be inserted into human cells will be those with a simple "always-on" type of regulation.

Given these constraints, the initial candidates for human gene therapy are the genes coding for the enzymes hypoxanthine-guanine phosphoribosyl transferase (HPRT), the absence of which results in Lesch-Nyhan disease, a lethal neurological disorder that can lead to uncontrollable self-mutilation; adenosine deaminase (ADA), the absence of which causes a severe combined immunodeficiency disease so that victims have to live in totally sterile environments; and purine nucleoside phosphorylase (PNP), the absence of which leads to another form of severe immunodeficiency disease. Approximately 200 new cases of Lesch-Nyhan disease are reported in the United States each

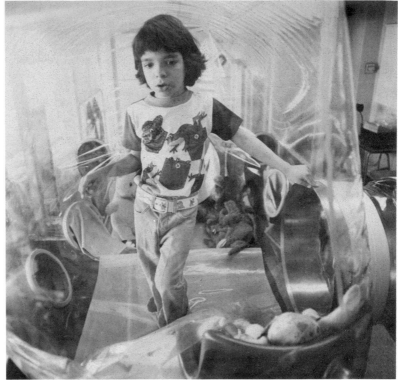

Baylor College of Medicine

Individuals lacking the enzyme adenosine deaminase (ADA) have such severely impaired immune systems that they must live in totally sterile environments to survive. For instance, David, the famous "Bubble Boy" (shown here at age 5), lived 12 years in isolated living quarters. By infecting the defective bone marrow cells of such patients with genetically engineered viruses containing the gene that codes for ADA, researchers hope to cure the disease.

year, making this the most common of the initial candidates for gene therapy. However, the neurological component of Lesch-Nyhan disease is caused by a lack of HPRT in the brain, and it is not known if supplying the enzyme from the bone marrow will overcome this deficit. The other two diseases are much rarer: only 40 to 50 cases of ADA deficiency and 9 cases of PNP deficiency are known worldwide.

The key step in the treatment of these diseases will be the insertion of genetically engineered copies of the respective genes into bone marrow cells removed from the patient's body. Researchers have been investigating several ways of introducing DNA into animal cells, including microinjection, chemically or electrically induced uptake, or fusion of the cells with vesicles containing the new DNA. But the most promising technique, and the one now being developed for human gene therapy, is the infection of the cells with genetically engineered retroviruses. As described in Chapter 3, retroviruses can insert a single copy of a DNA strand into the cells they infect. By attaching the appropriate regulatory signals to the inserted DNA, the gene can also be made to function within its new host. However, the position of insertion of the foreign gene into the host's DNA is random.

After the bone marrow cells have been transformed, they will be reimplanted into the patient. A limited number of studies have

THE GENETIC ENGINEERING OF HUMAN CELLS

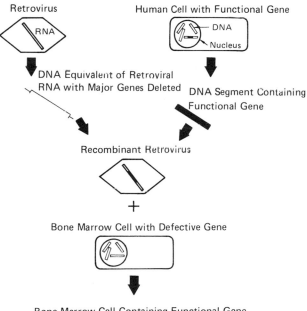

Upcoming attempts at human gene therapy will use retroviruses, infectious agents that have the ability to insert a single strand of DNA into the DNA of cells they infect. The genetic material of a retrovirus consists of RNA, which is enzymatically copied into its close chemical cousin DNA when the virus invades the cell. The major genes in a DNA copy of the retroviral RNA can be deleted and replaced with the desired gene from a human cell, along with the appropriate regulatory signals to ensure the expression of the gene. The bioengineered retroviruses can then be used to infect bone marrow cells withdrawn from a patient with a defective gene. The retroviruses insert the functional gene into a random location in the cells' DNA, and the transformed cells are reimplanted into the patient.

suggested that bone marrow cells that can produce HPRT and ADA have a growth advantage over bone marrow cells that cannot. If so, they will eventually come to predominate over a patient's defective bone marrow cells. If not, the defective bone marrow may have to be weakened or destroyed, through irradiation or other means, so that the transformed bone marrow cells can proliferate.

According to Anderson, several conditions must be met before such a procedure will be ethically permissible in human beings. The new genes must enter the proper cells and remain stable in those cells long enough to have the desired effect. They must also express their products at a level that will ameliorate the disease. Researchers are subjecting both of these conditions to rigorous study in tissue culture and laboratory animals to demonstrate their feasibility.

A final, more demanding condition is that the procedure not harm the cells to which it is applied or, by implication, the person receiving those cells. For example, a major concern is that the viral DNA used to transform bone marrow cells might naturally recombine with other pieces of DNA in the cell to form new infectious viruses, which could then spread to other cells. Researchers are looking for such recombinant viruses in tissue culture and laboratory animals to determine if this is possible. Other researchers are working to genetically engineer safeguards into the genetic material of retroviruses so that such recombinations cannot occur.

These conditions, which essentially amount to demands of delivery, expression, and safety, are no more than would be required of any new drug treatment or surgical procedure, and for a good reason. Somatic cell gene therapy differs little in its practical application from these more traditional treatments. "Somatic cell gene therapy is not fundamentally different from other kinds of medical care," says Leroy B. Walters, Jr., of Georgetown University's Center for Bioethics. "In particular, it is very similar to transplantation techniques, and especially to bone marrow transplantation techniques." Consequently, a consensus has been growing among those who have studied human gene therapy that it would be unethical to deny this treatment to desperately ill patients once the basic conditions of delivery, expression, and safety have been satisfied.

Nevertheless, a thorough review process has been set up to monitor the initial attempts at human gene therapy. After review by local Institutional Review Boards and Institutional Biosafety Committees, the research protocols for human gene therapy will have to be approved by a working group of NIH's Recombinant DNA Advisory Committee (RAC), by the committee itself, and by the director of NIH. (See

REGULATORY APPROVAL STEPS FOR HUMAN GENE THERAPY

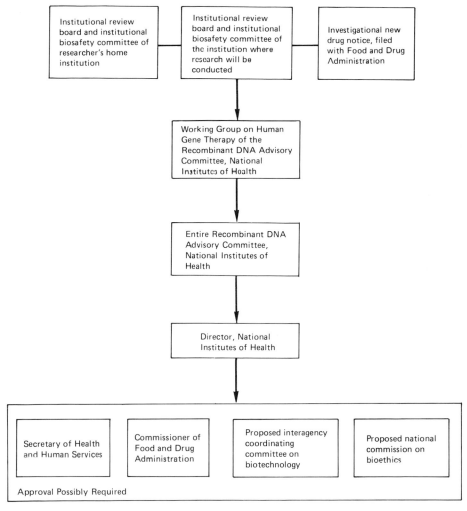

Experiments involving human gene therapy will require approval at several different levels before they can proceed. The review boards and biosafety committees of the researcher's home institution and the institution where the work is to be conducted must first approve the research protocol. The researcher must also file an investigational new drug notice (IND) with the Food and Drug Administration, although the agency does not have to approve the IND before the experiment can begin. The protocol then has to be approved at three separate levels within the National Institutes of Health. Finally, one or more other groups or individuals may have to approve the protocol, including the secretary of Health and Human Services, the commissioner of the Food and Drug Administration, the interagency coordinating committee proposed by the Cabinet Council Working Group on Biotechnology (discussed in Chapter 6), and the national commission on bioethics proposed by Senator Albert Gore, Jr.

Chapter 6 for a detailed discussion of the history and function of the RAC.) The Food and Drug Administration will regulate human gene therapy under the same regulations that it applies to clinical trials of any new drug or biologic (also discussed in Chapter 6).

As somatic cell gene therapy evolves and new procedures are developed, other concerns will come to the fore. For instance, it may someday be possible to conduct gene therapy on cells within the body as well as on cells withdrawn from the body. If genetically engineered retroviruses could be designed that home in on certain types of somatic cells when injected into the body, DNA could be delivered to specific body tissues. It appears, however, that a cell must be dividing for retroviral DNA to be incorporated into its chromosome, which would preclude the use of this technique for mature brain or nerve cells. Also, the introduction of viruses into the body would undoubtedly raise a host of additional questions about safety.

Germline Gene Therapy

Much more controversial than the replacement of a defective gene in somatic cells is the replacement of a defective gene in germline cells—the cells that contribute to the genetic heritage of offspring. In this case, gene therapy has the potential to affect not only the individual undergoing the treatment but his or her progeny as well. Germline gene therapy would change the genetic pool of the entire human species, and future generations would have to live with that change, for better or worse.

Germline gene therapy has been accomplished in laboratory animals. The mice described in Chapter 3 that were transformed by an inserted growth gene passed the gene on to their offspring, demonstrating that the gene had been inherited as a stable genetic trait.

But a number of technical difficulties make it extremely unlikely that germline gene therapy will be attempted in humans in the near future, if ever. First, the procedure has a very high failure rate. Most fertilized mouse eggs are so damaged by the microinjection and transfer that they never develop into live animals. Furthermore, a good laboratory can get the foreign gene into the germ line of only about 10 to 30 percent of the mice that are born. This degree of success occurs in mice that have been carefully inbred to give good results in this procedure. It would probably be even lower in genetically heterogeneous human cells.

A second major barrier is that none of the methods of inserting foreign DNA into cells offers any control over where the DNA will integrate into the chromosome. A foreign gene may be inserted into the middle of a critical gene in the cell, blocking that gene's function. The insertion of a foreign gene can also turn other genes in the cell on or off, causing metabolic imbalances that harm the cell. If this happens in a few of the bone marrow cells treated by somatic cell gene therapy, the consequences might go unnoticed. If it happens in a germline cell, the consequences are more likely to be severe.

The inability to control the chromosomal location of insertion points toward a much more fundamental problem that affects both somatic cell and germline gene therapy. The control of a gene in a cell depends on a number of regulatory factors, including position, and many of these regulatory influences are at this point not understood. It is impossible to insert the new gene into the exact position of the defective gene, because the defective gene is already there, and there are no techniques available for deleting or repairing a defective gene. Foreign genes have been genetically engineered to carry their own regulatory signals so that they are expressed in the cell. But these regulatory systems are crude compared with the precise regulatory systems of the cell. Some of the mice treated with growth hormone genes suffered from gigantism, so that parts of their bodies grew disproportionately large, and the metabolic imbalances caused by the inappropriate expression of growth hormone left nearly all the genetically engineered female mice sterile.

Because of these and other technical difficulties, there are no plans now being made to attempt germline gene therapy in human beings, and the prospects for any such attempts in at least the near future look dim.

Genetic Engineering to Enhance Human Traits

All the procedures discussed in the preceding sections are designed to insert into cells a normal gene corresponding to a defective gene. However, another kind of gene therapy—which Anderson feels is more properly termed genetic engineering—can also be imagined. This is genetic manipulation that seeks to insert genes into either somatic or germline cells in a way that alters or improves normal human attributes.

An example of such genetic engineering is the introduction into human cells of a gene that would produce elevated levels of growth hormone. This is not now possible with humans, but it is being attempted with livestock animals to increase their production of meat or milk. Given the present level of understanding, the physiological consequences of such genetic engineering cannot be predicted. A person who produces excess growth hormone would probably be taller, but the debilitating symptoms and disfiguration of people who naturally produce too much growth hormone, including a susceptibility to diabetes and heart disease and an appearance characteristic of gigantism, would probably also be present.

There are other examples of enhancement genetic engineering in which the issues are less clear-cut. Anderson points out the case of a gene that would produce more receptors for low-density lipoproteins on the surfaces of cells, reducing the level of cholesterol in the blood. Such a gene could bring the cholesterol level of people at the greatest risk of atherosclerosis down to its lower ranges. "This is an issue that doesn't need to be discussed in the immediate future," says Anderson. "But it is something that might very well come up in later years."

This ability to change specific physiological indexes should be sharply distinguished from the type of genetic engineering that has generated the most concern among the public, according to Anderson. This latter type of genetic engineering involves changing complex human traits—like intelligence, character, and physical appearance—that are shaped by a subtle interplay of many interacting genes and environmental influences. Such "eugenic" genetic engineering "really is a fantasy at the present time," says Anderson. "Any of these characteristics involves hundreds or thousands of genes interacting in completely unknown ways. How to be able to go in and insert one gene or two genes and in any way predictably change these enormously complex polygenic characteristics is totally unknown."

It is difficult to assess in scientific terms the likelihood of eugenic genetic engineering, because, as Anderson puts it, there simply isn't any science to discuss. But the ethical issues surrounding germline gene therapy or more straightforward forms of enhancement genetic engineering should not be slighted simply because these capabilities are not yet in hand. Technical advances are occurring at an increasing rate in molecular biology, and it is almost impossible to predict what eventually will or will not be doable.

Walters feels that the premature attempts at human gene therapy have given rise to "a very meaningful process of public reflection and discussion." In 1982 the President's Commission for the Study of Ethical Problems in Medicine and Biomedical and Behavioral Research released a report that clarified many of the basic issues surrounding the different types of human gene therapy, drawing on the expertise of not only scientists and physicians but philosophers, sociologists, and theologians as well. Since then several other hearings, symposia, and reports have elaborated on these issues, and they have received considerable attention from the more general media.

To continue this ongoing discussion, Senator Albert Gore, Jr., has proposed the creation of a national commission to monitor developments in biotechnology that affect human genetic engineering. The commission would be interdisciplinary and nonregulatory in nature, with the main goal of rendering advice and recommendations about the ethical implications of new capabilities. "We are at the present time woefully unprepared to grapple with the serious ethical choices with which the new technology will confront us," says Gore. "The very power to bring about so much good will also open the door to serious potential problems. If we are not careful, we may well cross the line separating the two. Knowing where that line exists is the challenge that we face."

Additional Readings

W. French Anderson. 1984. "Prospects for Human Gene Therapy." *Science* 226 (October 26): 401-409.

Yvonne Baskin. 1984. *The Gene Doctors*. New York: William Morrow.

Office of Technology Assessment. 1984. *Human Gene Therapy—Background Paper*. Washington, D.C.: U.S. Government Printing Office.

President's Commission for the Study of Ethical Problems in Medicine and Biomedical and Behavioral Research. 1982. *Splicing Life: A Report on the Social and Ethical Issues of Genetic Engineering with Human Beings*. Washington, D.C.: U.S. Government Printing Office.

5

The Release of Genetically Engineered Organisms into the Environment

A LITTLE MORE THAN A DECADE after the basic techniques of genetic engineering were pioneered, biotechnology is about to enter an entirely new domain. Over the next few years the first organisms genetically engineered for use in the environment will be field tested and put to work. The first of these organisms to reach the marketplace will probably have agricultural applications. Genetically engineered microorganisms will be used to control insect pests, to fix nitrogen, and to reduce frost damage. Bioengineered crops may be hardier, resistant to different kinds of pesticides, and more productive. (Chapter 3 discusses potential agricultural applications in detail.) Genetically engineered organisms could also find nonagricultural uses in wastewater treatment facilities, in mining operations, and in oil wells. (Some of these possible applications are described in Chapter 2.)

In addition to offering dramatic new capabilities, the use of genetically engineered organisms in the environment will raise a host of new concerns. Most natural ecosystems are exceedingly complex assemblages of many different organisms and abiotic influences, and many of the relationships among an ecosystem's components are still poorly understood. A genetically engineered organism introduced into an

---

This chapter includes material from the presentations by Martin Alexander and Daniel Nathans at the symposium.

ecosystem will therefore have the potential of affecting it in unanticipated, and possibly detrimental, ways. Moreover, unlike air or water
pollution, which tends to dissipate over time, organisms have the
capacity to reproduce and spread, magnifying any problem that does
arise.

The history of conventional plant and animal breeding suggests that
the likelihood of a problem is very low. Plant and animal breeders have
been creating new crop varieties and livestock breeds for many centuries without posing untoward risks to the environment.

On the other hand, plants, animals, and microorganisms introduced
into new locales from other parts of the world have caused major and
lasting environmental damage. A fungus introduced into America from
Asia killed almost all of North America's majestic chestnut trees.
Another fungus has eliminated most Dutch elm trees from the eastern
United States. A virus introduced into Australia almost completely
annihilated that continent's rabbit population. Over half the insect
pests in the United States today come from abroad. Kudzu and hydrilla
are two examples of weeds introduced into the United States that have
caused monumental problems. Similarly, starlings, house sparrows,
and gypsy moths are all introduced animals that America would
almost surely be better off without.

By the same token, most of America's major crops, including soy
beans, wheat, and rice, are not indigenous to this continent. Most
livestock breeds and many poultry breeds have their roots in Asia,
Africa, or Europe. Farmers and gardeners even apply *Rhizobium*
bacteria from other parts of the world to their fields and plots to boost
yields.

Such arguments by analogy are valuable reminders of biotechnology's potential to do great good or great harm, but they leave much
unexplored when it comes to examining the release of genetically
engineered organisms into the environment. Genetic engineering offers capabilities that range far beyond those of conventional breeding
programs. Understanding the issue therefore requires a much more
detailed examination of the specific factors involved when a new
organism is introduced into the environment.

The Components of Environmental Risk

According to Martin Alexander of Cornell University, five independent factors come into play in determining what effect a genetically
engineered organism will have on other organisms. These are (1)
whether the organism is released into the environment, (2) whether it

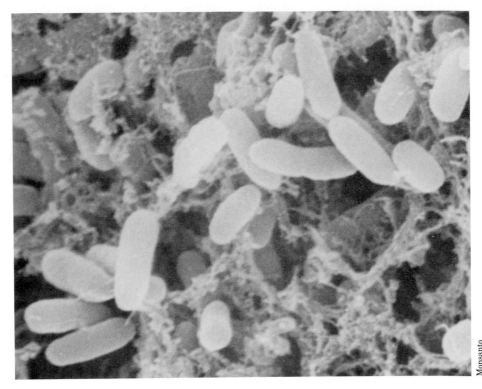

Monsanto

Common soil bacteria that have been genetically engineered to contain the gene for a pesticidal toxin colonize the roots of a corn plant. Such genetically engineered organisms must be thoroughly tested to ensure that they will not have unintended detrimental effects on ecosystems into which they are introduced.

survives, (3) whether it multiplies, (4) whether it moves to an area where it can have an effect, and (5) what that effect actually is. In addition, DNA may be transferred between organisms in the environment, either sexually or asexually, and this must also be taken into account in calculating the risk posed by a genetically engineered organism.

Release

Obviously, an organism must first enter the environment to cause harm. Most of the organisms that have been genetically engineered to date have been designed for laboratory research or fermentation processes, and so long as they stay within the fermentor they pose no risk to the environment. Moreover, these organisms have generally been genetically crippled to make it very difficult for them to survive outside the flask or fermentor.

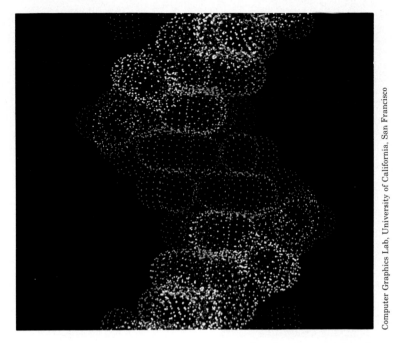

Computer Graphics Lab, University of California, San Francisco

A color-coded computer model of DNA—the molecule that carries the genetic blueprint of every living organism.

du Pont

Fragments of DNA radiate under ultraviolet light. These DNA pieces can be inserted into microbes to generate large quantities of insulin and other useful products.

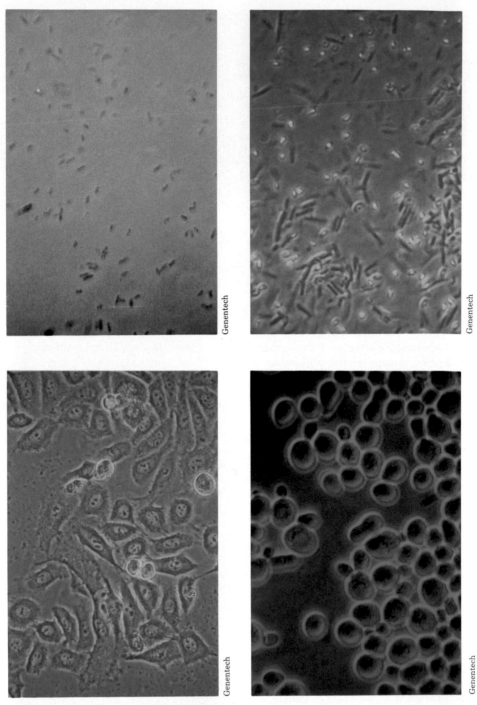

Common hosts for inserted DNA (clockwise from top left): *Escherichia coli*, *Bacillus*, yeast, and mammalian cells.

National Institutes of Health

Flasks filled with microbes that have been genetically engineered to produce interferon, a compound that could prove useful in checking some cancers and certain types of infections.

Monsanto

Fermentors ranging in size from this laboratory apparatus to vessels containing many thousands of gallons are used to grow large quantities of genetically engineered microbes.

Eli Lilly

Ultrafiltration—passing proteins through narrow hollow fibers—can be used to purify insulin and other proteins generated by genetically engineered microbes.

Eli Lilly

A drop of human insulin made by genetically engineered bacteria.

Tumor and spleen cell hybrids are cultured to generate large quantities of monoclonal antibodies, compounds valuable in scientific research and in the diagnosis and treatment of disease.

National Institutes of Health

Centocor

A researcher tests the ability of monoclonal antibodies to latch onto specific antigens.

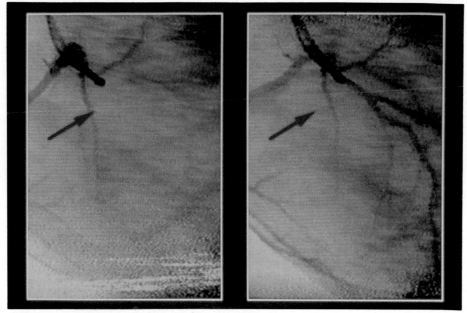

Genentech

Blood flow is restored (right) to a blocked artery (left) after an injection of tissue-type plasminogen activator—a genetically engineered compound that dissolves blood clots. This compound can effectively treat heart attack victims.

Steven E. Lindow, University of California, Berkeley

Frost damaged the potato plant on the left, but the plant on the right, from which the bacteria that initiate frost formation had been removed, emerged unscathed. If genetically engineered bacteria lacking the ability to cause ice nucleation could be sprayed on plants, frost damage could be minimized.

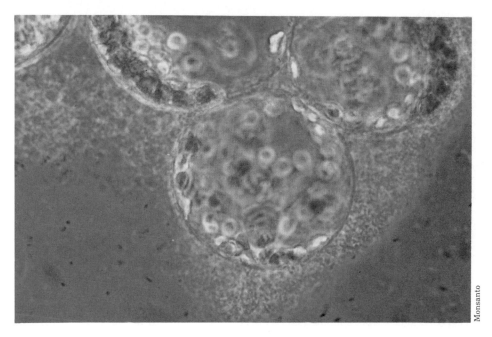

Plant cells surrounded by genetically engineered soil bacteria. Scientists use these bacteria to introduce desirable genes into plants, with the eventual goal of improving agricultural crops.

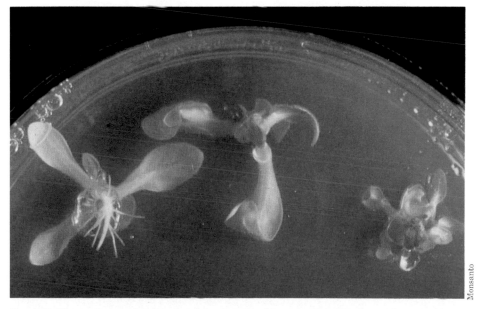

Genetically engineered petunia plants grown from single cells through exposure to hormones and nutrients.

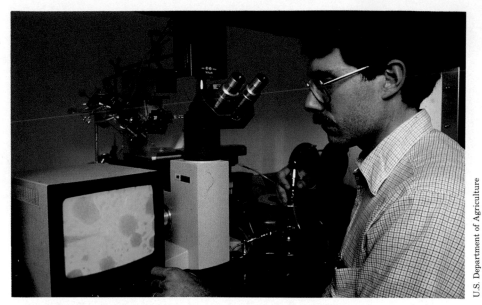

U.S. Department of Agriculture

With the aid of a television monitor, a researcher guides a fine glass needle (on left of screen) containing DNA into the nucleus of a cell. This technique is used to introduce desirable genes into plants and animals.

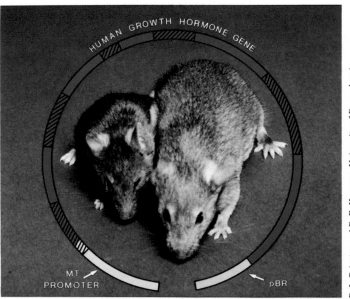

R. L. Brinster and R. E. Hammer, University of Pennsylvania

The mouse on the right is twice the size of its sibling because of a growth-promoting gene that was inserted into its cells. The ring surrounding the mice depicts the genetic material introduced into the larger mouse's cells.

Genetically engineered organisms undergo initial testing in isolated growth chambers such as the one shown here. But growth chambers cannot re-create the full complexity of a natural ecosystem, and at some point small-scale field testing becomes necessary.

A few bioengineered organisms designed for use in the environment, including microbial pesticides and genetically engineered crops, have already been produced and studied within laboratories, growth chambers, and greenhouses. Experiments involving these organisms carried out with public funds are required to adhere to the appropriate containment procedures specified in NIH's *Guidelines for Research Involving Recombinant DNA Molecules* (the history of these guidelines appears in Chapter 6). These procedures are meant to ensure that the likelihood of a potentially hazardous organism escaping into the environment remains very low. For the purposes of calculating a probability, however, it should be remembered that accidental releases of hazardous organisms from research facilities have occurred in the past. As Alexander says, "no tank never leaks." Even stringent efforts at containment are only "reducing the probability of an accident, not converting it to zero probability."

Regardless of the developmental research involved, an organism genetically engineered for environmental applications will eventually be ready for small-scale field testing. At that point, the probability of release becomes one.

Survival

Once an organism has been released into the environment, it must survive to have an effect. As Alexander points out, predicting the survival of an introduced organism is one of the most difficult problems in ecology. "Natural ecosystems have what are known as homeostatic mechanisms," he explains. "There are a variety of interactions among plants, animals, and microorganisms that tend to keep in check the rare species, eliminate alien species, and prevent the dominant species from overexploiting the environment. If we go into a particular area, we see the same plants, and an occasional introduction of a new organism will not result in its establishment in the field.

"Homeostasis is effective in eliminating aliens, but it is not always wholly successful. In plant ecology, animal ecology, and microbial ecology, it is known that an introduced organism does occasionally survive."

Alexander's own research and the studies of other microbial ecologists have revealed many instances of microorganisms from foreign locations that survive for days, weeks, months, and even years when introduced into new environments. The past history of introduced plants, animals, and microorganisms that have done great environmental damage, while admittedly worst-case examples, also indicates that some percentage of introduced organisms will survive.

The problem lies in determining which organisms will survive and why. "The general feeling is that homeostasis will eliminate nearly all, but not all, species," says Alexander. "Even given the successful establishment of an organism—plant, animal, or microorganism—we cannot explain why that one was successful and many others failed." Thus there is considerable uncertainty surrounding this component of environmental release. However, if a genetically engineered organism is to have its intended effect, it must survive for at least some period of time.

Multiplication

Most genetically engineered organisms must also multiply if they are to have an effect. In general, the number of organisms originally released will be too small to do much harm. "The organism must reach a population density high enough to upset other organisms—either a microorganism upsetting host plants or animals, a plant that becomes a weed because it is abundant, or an animal that disturbs its natural environment," says Alexander.

The determinants of successful multiplication are in most cases as unknown as those of survival. Says Alexander, "Except for pathogens, we do not know under what conditions almost any microorganism can multiply in nature. We cannot predict the organisms that will multiply. This is true of most species of plants as well as of microorganisms. All we have are instances with particular economically important species."

One factor that may work against the survival of a genetically engineered organism is that the organism contains extra DNA, which diverts part of its metabolic energy from the pursuit of survival and multiplication to the production of agriculturally important proteins. In this way, the genetically engineered organism is at a competitive disadvantage with organisms that do not bear the burden of extra DNA. However, as Alexander points out, the ecological consequences of extra DNA need not be wholly negative. "If the acquisition of one characteristic results in an ecological advantage, then the organism may be able to overcome one of the environmental barriers to its establishment. Unfortunately, at this time, we can't tell whether additional DNA that is disadvantageous in one way will also be advantageous in another way."

Furthermore, one could argue that it is unlikely that a genetically engineered organism would acquire a newfound persistence, as in the case of a weed or rampant pathogen, because many interacting genes are needed to generate such characteristics. Here too, however, counterexamples can be cited in which small genetic alterations lead to major changes in an organism's behavior. Although not related to genetic engineering, slight changes in the antigenicity of an influenza virus can lead to reduced immunity among humans and a greater severity of the disease. Similarly, the genetically straightforward formation of a capsule around some bacteria can make them resistant to normal human and animal defenses. In agriculture, the addition of a gene for resistance to pests or herbicides or the acquisition of genes for more efficient photosynthesis could give plants an edge over their nonengineered competitors.

Dispersal

An organism usually will not cause harmful effects in the area where it is released—a farmer's field, a waste dump, a tailings pond. Instead, it must move to an area where it encounters organisms susceptible to its effects. Thus the greater the range of a genetically engineered organism, the greater its chances of causing a problem.

The dispersal of some organisms has been studied extensively, according to Alexander, but much less is known about other organisms. Yet it is crucial that the range of an organism be determined before its release, since genetically engineered organisms will generally have been designed to survive and multiply in the environment.

Effects

Finally, the effects of a released organism on other living things in the environment—microbes, plants, animals, and humans—must be calculated. In some cases these effects, if they occur, will be obvious; in others they will be indirect and subtle. To take just one example from traditional plant breeding, a specific cultivar of potato had to be removed from supermarket shelves because it was found capable of producing hazardous levels of toxins under certain conditions of stress.

Such unintended effects may be less likely to occur with recombinant DNA techniques than with traditional plant and animal breeding, since the genes and metabolic pathways to be altered are likely to be more fully characterized with recombinant DNA. Nevertheless, the search for effects will be difficult in many cases, because the interrelations among organisms in the environment are often poorly characterized.

The Transfer of Genetic Information

An additional complication in calculating the risks of environmental release is that organisms in the environment can transfer DNA to other organisms through a variety of means. If a genetically engineered organism transfers its new traits to another organism, the string of risk factors, from survival to effects, must be calculated anew.

The most common means of genetic transfer among plants and animals is sexual recombination. For instance, pollen from a genetically engineered crop could fertilize the seeds of a similar but nonengineered crop. Much more worrisome is the possibility of crosses between crops and related noncrops or weeds. Just as plant breeders transfer traits from cross-fertile weeds into agriculturally important plants through crossbreeding followed by successive backcrosses—a process known as introgression—so a gene from a crop could be transferred to a related weed through natural introgression. In fact, this seems to have occurred several times in the past, such as in the relationship between weedy Johnson grass and sorghum.

There are several forces that will probably keep this from being a serious problem in biotechnology. For one thing, only a few species of weeds cross in nature with major crop plants. Nor is it clear just how much advantage any weedy species has ever gained from natural introgression. If problems were anticipated, geographic limitations on the use of genetically engineered crops could be imposed. Finally, natural introgression is a concern with the introduction of any new plant or animal breed, not only with those that will be produced through biotechnology.

Plants are not known to transfer DNA between one another through nonsexual means, and such transfers appear to be rare among animals (viruses are possible intermediaries of nonsexual genetic transfer). However, microorganisms exchange genetic material nonsexually in several different ways. Such exchanges of DNA have been known to transfer traits like resistance to antibiotics among microorganisms in laboratory and hospital settings. But it is not known if such transfers among microorganisms occur in natural environments, according to Alexander, and their impact of such transfers on the risk of environmental release is likewise unknown.

Risks and Uncertainties

The probability that a genetically engineered organism will have a detrimental effect on the environment is the product of the five factors discussed above: release, survival, multiplication, dispersal, and effects. (The last four factors also come into play for an organism that acquires foreign DNA from a genetically engineered organism.) Since the probabilities associated with one or more of these factors are likely to be small, the overall probability of a harmful effect is likely to be very small. But a low probability is not a zero probability. And, as Alexander points out, "the consequences of this low-probability event could be very significant."

The uncertainties surrounding each of the six components of environmental risk make it impossible to calculate precisely how small the risk is. Claims of zero risk or great risk are therefore inappropriate, according to Alexander, and merely muddy the debate surrounding the issue. Furthermore, the uncertainties will loom larger as more and more organisms are altered, as the number and kind of introduced genes grow, and as genetically engineered organisms are released into a wider range of environments. "The degree of uncertainty is too large for me, as an ecologist, to feel particularly comfortable with," concludes Alexander.

Research and Regulation

The way to reduce the level of uncertainty now associated with environmental release is through research into the interactions of organisms with their surroundings. Indeed, it is highly desirable that this research be done before the range of biotechnology's applications in the environment begins to expand, but very little of this research is now being supported by federal regulatory or research agencies, Alexander says.

Researchers should concentrate on several key points, according to Alexander. Most important, the specific factors that contribute to the probabilities associated with each of the six components of environmental risk should be identified. This would help the industry choose the organisms that it should use and rule out those that it should avoid. It could also help in fashioning debilitated organisms that would not survive, multiply, or disperse once their intended purpose was complete. The identification of these traits would facilitate the testing needed to evaluate environmental risk. It would also help regulators decide which organisms need extensive testing before approval and which need little or no testing.

Several important technical and methodological issues should also be addressed. For instance, ways need to be developed to label a genetically engineered organism so that its fate can be monitored in the field. This would greatly simplify studies of an organism's escape, survival, multiplication, and dispersal and may even help in tracking the movement of DNA among organisms in the environment. Several immunologic and genetic techniques have been adapted for labeling purposes, but they require further development.

Research along these lines would reduce many of the uncertainties surrounding environmental release, but it cannot eliminate them. Ecology is not such an exact science as to lend itself to infallible predictions. As Daniel Nathans of the Johns Hopkins University School of Medicine says, "It is difficult to envision how one will get the knowledge to tell you in a concrete way whether transfer of a particular gene into a particular organism is 20 years from now going to cause an ecological disaster. We will never do the experiment if you require that question to be answered in a scientifically acceptable way. . . . We are left with reasoned, conservative judgments of people in the field, and we are also left with very carefully controlled, step-by-step experiments, in which appropriate measurements are made."

The difficult task of balancing the remaining uncertainties against the undeniable benefits of biotechnology falls most immediately to the federal regulatory agencies that oversee genetic engineering and its application. As discussed in Chapter 6, industry representatives and government officials agree that the regulations established by these agencies will be a critical and often indispensable factor in the industry's development. According to Alexander, these regulations will reduce the possibility of an ecological upset. They will also ease the public's fears about the new technology. They will help the industry to get liability insurance at reasonable rates. And they will reduce the backlash when a problem does occur or when a problem that arises is mistakenly attributed to industry.

The pursuit of the environmental applications of genetic engineering therefore involves three overlapping fronts: the development of the organisms, research on the interactions of the organisms with the environment, and regulation of the organisms' development and application. By moving forward on these three fronts simultaneously, it should be possible to reap the benefits of biotechnology while holding the risk to the environment at a minimum. "If one has a good base of scientific information and a reasonable testing system, then I think that much of the residual degree of uncertainty can easily be answered by a very modest regulatory program," says Alexander. "But we should have a regulatory system in place, a regulatory system that will reduce the likelihood of a problem arising, and a significant amount of research to find out where the issues are."

Additional Readings

Martin Alexander. 1985. "Ecological Consequences: Reducing the Uncertainties." *Issues in Science and Technology* 1(Spring):57-68.

Winston J. Brill. 1985. "Safety Concerns and Genetic Engineering in Agriculture." *Science* 227(January 25):381-384. [See also the responses to this article in *Science* 229(July 12, 1985):111-118.]

Bernard Dixon. 1985. *Engineered Organisms in the Environment: Scientific Issues.* Washington, D.C.: American Society for Microbiology.

Ecosystems Research Center. 1985. *Potential Impacts of Environmental Release of Biotechnology Products: Assessment, Regulation, and Research Needs.* Ithaca, N.Y.: Ecosystems Research Center, Cornell University.

Holly Hauptli, Nanette Newell, and Robert M. Goodman. 1985. "Genetically Engineered Plants: Environmental Issues." *Bio/Technology* 3(May):437-442.

Albert H. Teich, Morris A. Levin, and Jill H. Pace, eds. 1985. *Biotechnology and the Environment: Risk and Regulation.* Washington, D.C.: American Association for the Advancement of Science.

6

Governmental Regulation of Biotechnology

THE USE OF RECOMBINANT DNA has been subject to public overview almost since the technology's inception. In 1973, as word of the exciting new ability to join DNA from different organisms began spreading through the scientific community, a group of scientists involved with the research sent a letter to Philip Handler, then president of the National Academy of Sciences, pointing out that the new capability presented possible hazards as well as great scientific promise. Out of that letter arose a chain of events that led to two of the most important events in the early history of genetic engineering: a voluntary moratorium on certain types of recombinant DNA experiments deemed particularly hazardous, and the International Conference on Recombinant DNA Molecules, which was held at the Asilomar Conference Center in Pacific Grove, California, February 24-27, 1975.

Although primarily a scientific meeting, Asilomar was marked by a debate that had a prominent public policy component. On one side were those who held that research with recombinant DNA should proceed

This chapter includes material from the presentations by Alexander Rich, Albert Gore, Jr., Richard J. Mahoney, William B. Ruckelshaus, Joseph G. Perpich, Bernadine Healy, William J. Gartland, Jr., Harry M. Meyer, Jr., Brian Cunningham, John A. Moore, Geoffrey M. Karny, Orville G. Bentley, Irving S. Johnson, Robert P. Nicholas, Thomas O. McGarity, and Zsolt Harsanyi at the symposium.

Andrew A. Stern

Participants at the 1975 Asilomar conference. The legacy of Asilomar remains a powerful influence on biotechnology even 10 years after the conference occurred.

untrammeled by guidelines or regulations. On the other were those who believed that the potential dangers demanded restrictions, or at least that self-imposed guidelines were far preferable to regulations imposed from outside the scientific community. In the end the latter group prevailed, and a statement of principles outlining a proposed set of standards for recombinant DNA research was drafted.

The day after the conference a committee of scientists appointed by the National Institutes of Health, now known as the Recombinant DNA Advisory Committee (RAC), began converting the statement of principles into formal guidelines. Issued in June 1976, the *Guidelines for Research Involving Recombinant DNA Molecules* assigned different types of experiments to different categories of risk. Certain experiments were prohibited outright. Others had to be conducted using various levels of physical and biological containment, involving special laboratory equipment and procedures or attenuated hosts (most commonly the K-12 strain of the bacterium *Escherichia coli*).

Ten years after Asilomar, impressions of the meeting and its aftermath still differ. "For many people in the nonscientific community, it is viewed as an act of scientific statesmanship at a high level," says Alexander Rich of the Massachusetts Institute of Technology. "The other side of the coin, for many members of the scientific community, is that it was a mistake, that the scientists overreacted, that had they thought more carefully and looked at the available evidence they would have understood that the hypothetical risks were in fact imaginary."

As experience with recombinant DNA accumulated, it became clear that many of the risks associated with the research either did not exist or were initially overestimated. "We have had between 100,000 and a million experiments worldwide in recombinant DNA activities without, as one scientist has said, a sniffle," Rich points out. "The scare scenarios were in fact erroneous." This reevaluation of risk has led to successive revisions of the NIH guidelines. In 1978, just one year after the National Academy of Sciences Forum on Research with Recombinant DNA, where the hypothetical risks were a great concern, the ban on the forbidden experiments was lifted, although several still require the approval of the RAC and the director of NIH. Today nearly 90 percent of the experiments involving recombinant DNA are exempt from the guidelines.

Nevertheless, new regulatory concerns have emerged. As biotechnology begins to generate products for the marketplace, federal regulatory agencies have channeled those products into existing regulatory processes. The Food and Drug Administration traditionally approves new drugs and biologics before they can be marketed. The Environmental Protection Agency regulates certain microorganisms to be used in the environment. The U.S. Department of Agriculture oversees the importation and interstate commerce of agriculturally important plants, animals, and microbes. At the same time, research has progressed so quickly that previously prohibited experiments for which concern about safety or ethics still exists, such as the release of genetically engineered organisms into the environment or human gene therapy, are ready to begin.

This rapid evolution of biotechnology has left the government with several distinct goals. It has the responsibility to protect human health and the environment from any risks posed by biotechnology, even though the extent of possible risks is conjectural. Simultaneously, it

Paul Conklin for the Academy Forum

Demonstrators at the 1977 National Academy of Sciences Forum on Research with Recombinant DNA. The early fears about the safety of recombinant DNA research have largely quieted as experience with the new techniques has accumulated.

has an interest in seeing the biotechnology industry prosper, not only because of the specific products that will emerge from the industry but because of the broad economic benefits to be expected of a vibrant and expanding industrial sector. "A central purpose of any governmental effort in this area must be to encourage and facilitate the growth of biotechnology research, development, and implementation," says Senator Albert Gore, Jr. "We are already beginning to face serious competition from other nations for leadership in biotechnology. I, for one, am determined that we remain in the forefront. The government should help guide biotechnology, but it must not control it."

Industry leaders share these convictions, pointing out that the United States' current lead in converting the results of basic biomedical research into commercial products is fragile. (Chapter 10 discusses international competition in biotechnology in more detail.) They assert that if the government imposes burdensome regulations on biotechnology, new products will take longer to reach the marketplace. Biotechnology firms in countries with less encumbering regulations could then catch up with and surpass their American counterparts, securing patents and market presences that would thereafter be denied to American firms. "It would be a tragedy of enormous proportions if the promise of biotechnology were unfulfilled," contends Richard J. Mahoney of Monsanto. "I, for one, don't relish the prospect of American farmers buying their genetically engineered wheat and corn seeds from Japan or Europe or elsewhere, or turning elsewhere for their latest miracle drugs. . . . But mark my words, there is a possibility that if regulatory delays prevent the timely development of these products in the United States, we will lose our lead. America has pioneered a truly great technology, and we deserve some of the economic benefits that will flow from it."

At the same time, industry acknowledges the many benefits to be gained from a stable and sound regulatory regime. For one thing, strong governmental oversight can help build the public trust that is essential for industries perceived as potentially hazardous to progress. According to William B. Ruckelshaus, former head of the Environmental Protection Agency, "If there is one thing that we learned from the recent upheavals at the EPA, it is that when the agency charged by society with protecting the public health and the environment comes apart, to the point that society no longer really trusts that institution, then those dependent on those decisions for the marketing of their products are in more trouble than anyone else."

The biotechnology industry is in a position somewhat similar to that of the scientists who gathered at Asilomar. It realizes that regulations emerging from a cooperative effort between industry and government will be preferable to regulations in which the industry has no say. "Industry has a vested interest in sensible, science-based regulation," says Mahoney. "It seems to me that business has two choices: become a partner in developing the guidelines that will ensure adequate protection, or be an adversary. In the latter case, regulations will emerge just as surely, but they will make our jobs a lot more difficult."

The government therefore faces the difficult task of coordinating its charge to protect human health and the environment with its desire to see the biotechnology industry thrive. It is a delicate balance, and errors in either direction could upset the competitive advantage that American biotechnology firms now enjoy. "If the government doesn't do its job, it is unlikely that the United States will stay ahead either in the research or the commercialization of this new technology," says Ruckelshaus. "There are two ways in which that could happen. One is that the government would do too much; there would be very heavy-handed regulation, needless time-delaying, red tape, nitpicking, an unwillingness on the part of the government ever to make a decision. If there is too much regulation, we could stifle this emerging technology. On the other side, that could also happen if there is too little regulation, or if it is unwise or sloppy or not open. [That] could just as certainly doom our efforts to take advantage of our current lead."

"This may be the last chance to do it right the first time," Ruckelshaus continues. "There is the potential for a cooperative effort on the part of the government, industry, academia, and public interest groups to move all of this forward in a way that is consistent with the protection of the environment and the commercialization of these products."

The Strengths and Weaknesses of the NIH Guidelines

One of the most remarkable features of the NIH guidelines is the pervasive influence they have had even though they are merely guidelines. Technically, they apply only to institutions that receive federal funds, and the penalty for violating the guidelines cannot extend beyond withdrawal of those funds. However, other governmental agencies in addition to NIH have also required that the guidelines be followed, and several states and localities have established the guidelines by law. In addition, private companies that work with recombinant DNA have adopted the guidelines and have submitted research proposals to the RAC. Even many foreign countries conducting recombinant DNA research have adopted slightly modified versions of the guidelines.

The RAC's distinguished members, its sophisticated deliberations, and the widespread influence of its guidelines have made it a respected central clearinghouse for discussion of the scientific and social issues surrounding biotechnology. The quasi-regulatory nature of the guidelines has also given them flexibility, allowing them to be revised in light of constantly accumulating scientific information. When gaps in expertise have been identified in the RAC, it has accepted additional members or enlisted outside consultants. Representatives of industry and government alike have praised its past performance and have urged that it continue in its role. "It is a testament to the NIH and to the scientific community for maintaining and nurturing this system over the past 10 years, and to a lot of very gifted people working on the RAC to develop and revise the guidelines," says Joseph G. Perpich of Meloy Laboratories, Inc. "We are here today talking about promise and products because of them."

At the same time, several limitations of the NIH guidelines and the RAC as they are presently organized have become apparent. First, the NIH guidelines do not have the force of law, and compliance with them by private companies is voluntary. Recently several companies have bypassed the RAC entirely and have submitted research proposals involving recombinant DNA to federal agencies with specific jurisdiction over anticipated products.

Second, the NIH guidelines, because of their history and evolution, inevitably focus more on research than on commercial development. This has led to questions from both inside and outside the RAC about the propriety of its continuing to oversee commercial developments in biotechnology. "If there is an inadequacy in the current federal structure, it is in the availability of a scientific review mechanism that can deal with the broad range of commercial products now emerging and on the horizon," says Bernadine Healy, formerly of the White House Office of Science and Technology Policy. "The NIH RAC is neither equipped nor desirous of taking on that role on a government-wide basis. The NIH is oriented toward basic biomedical research and not toward commercial scale-up or engineering. It has limited environmental and ecological expertise, and it does not want to take on that expanded chore."

The RAC's apparent jurisdiction over some commercial biotechnology products has also brought it into potential conflict with other federal agencies. Both the RAC and the Food and Drug Administration claim authority over clinical trials of human gene therapy. The Environmental Protection Agency and the U.S. Department of Agriculture both assert that they should oversee the use in the environment of certain genetically engineered microorganisms. "For commercial products—foods, drugs, and chemicals—there is a clearly established statutory responsibility in several other agencies outside of the NIH," says Healy.

Furthermore, the RAC only oversees research involving recombinant DNA. Genetic techniques such as cell fusion or mutagenesis are not part of its charter.

Finally, the RAC meets only four times a year, raising concerns that it might be unable to handle the full range of proposals that may need attention in the future.

The problems with the RAC's quasi-regulatory status became apparent during the initial wrangling over one of today's most pressing regulatory issues—the release of certain genetically engineered microorganisms into the environment. (For a discussion of the scientific issues surrounding environmental release, see Chapter 5.) The NIH guidelines require that any experiment involving the environmental release of genetically engineered organisms first be approved by both the RAC and the director of NIH. Between 1981 and 1983 the RAC and the NIH director approved three such experiments. The first was a field test of corn modified by recombinant DNA techniques to contain other corn gene sequences, but the experiment was never conducted because the corn turned out not to be ready to test. The second was a field test of modified tobacco and tomato plants, but this experiment was also scuttled because of technical problems. The third was a field test of bacteria that had been genetically modified to reduce frost damage in plants. (The principle behind this experiment is described in Chapter 3.)

In September 1983 a lawsuit was filed against NIH by the Foundation on Economic Trends, a public interest group, alleging that in approving these tests it had not complied with the National Environmental Protection Act, which requires that federal agencies prepare environmental impact statements for "major Federal actions significantly affecting the quality of the human environment." In response, U.S. District Court Judge John Sirica issued a preliminary injunction prohibiting the field test of the bacteria and prohibiting NIH from approving any further environmental releases of genetically engineered organisms. NIH appealed the order, and on February 24, 1985, the U.S. Court of Appeals for the District of Columbia ruled that NIH did not have to file an environmental impact statement but that it did have to prepare a less extensive environmental assessment. In the meantime, the Environmental Protection Agency stated that it, too, had authority over the environmental release of the genetically engineered microorganism in question. The lawsuit appeared to contribute to a renewed concern in Congress over genetic engineering: several legislators questioned whether new laws were needed to regulate forthcoming applications of biotechnology.

To those people who have followed genetic engineering since its origins, the recent furor over environmental release has a familiar ring. "The discussions about the possible hazards of releasing engineered organisms are reminiscent of the situation that existed before the Asilomar conference, when many different hazard scenarios were being raised," says William J. Gartland, Jr., of NIH's Office of Recombinant DNA Activities. "I think the issues here, though, are going to be much more complex than the issues that the Asilomar conference had to deal with. Asilomar was largely concerned with biomedical research, and it was largely concerned with practically one organism, namely *E. coli* K-12. I think, at the outset of this new phase of research with deliberate release, that there will probably be dozens of organisms that will be proposed to be released in dozens of different settings. It will be a much more complex issue to deal with."

The Roles of Other Federal Agencies in Regulation

The positions taken by federal agencies other than NIH will be crucial in determining the future course of the RAC and the overall regulation of biotechnology. These positions are still in the process of being established, and they are bound to change as the technology and the regulatory climate evolve. But a broad statutory and regulatory framework for biotechnology already exists, and it is this framework that will largely determine the future regulation of the industry.

The Food and Drug Administration

The Food and Drug Administration (FDA) will regulate applications of genetic engineering primarily under the Food, Drug and Cosmetic Act and the Public Health Service Act. These statutes give FDA authority over human and animal drugs, human biologics, food and color additives, and medical devices (including in vitro diagnostic tests that employ monoclonal antibodies), among other substances. Before a manufacturer begins to market a new drug or biologic, it must prove to the FDA through a variety of means, including clinical tests on humans, that the substance is "safe and effective." In turn, a "new" drug is defined as one that is not yet recognized by qualified scientific experts as safe and effective for its proposed use.

The FDA has traditionally allowed manufacturers to use certain abbreviated approval processes for products identical to already approved or existing substances that were manufactured by identical techniques. But it has decided that at least for the time being it will treat all drugs and biologics derived from methods involving recombinant DNA as new, requiring them to undergo the entire approval process. The reason for this caution is concern over the possibility of undetected or novel contaminants in the product—for instance, the endotoxins produced by *E. coli* as part of its metabolic processes—or the possibility of genetic instability in a recombinant organism.

The approval process for new drugs begins with the submission by a manufacturer of an investigational new drug notice (IND). The IND contains information about the structural composition of the drug, the manufacturing process, the results of animal testing, the plans for clinical trials in humans, the consent forms to be used with human subjects, the background of the investigators, and other data required to demonstrate that the drug will be safe for human testing. Unless the FDA disapproves the IND, the clinical investigations can begin. These are also closely monitored by the FDA.

The second major step in the approval process is the submission of a new drug application (NDA). The NDA contains a full report of the results of the clinical trials in humans, a statement of the drug's quantitative composition, and a description of the methods and controls used to manufacture, process, and package the drug. If changes are made to an IND or NDA before or after its approval, amendments must be submitted and approved by the FDA, some of which may require additional clinical testing.

The approval process is somewhat different for human biologics, which the law defines as any "virus, therapeutic serum, toxin, antitoxin, vaccine, blood, blood component or derivative, allergenic product or analogous product . . . applicable to the prevention, treatment, or cure of diseases or injuries." The IND phase of the application is similar, but the successful completion of clinical testing results in a license rather than in the approval of an NDA. The biologic must meet certain standards of safety, purity, potency, and efficacy; its manufacturing facilities must undergo prelicense inspections (and are themselves licensed); and the licensed products are subject to lot-by-lot testing by the FDA.

The procedures for the approval of a new drug or food additive for animals are similar to those for drugs or food additives for humans. In addition, drugs for use in animals must not leave unsafe residues in the edible tissues of food products.

New medical devices must also be approved by the FDA before they can be marketed. However, the law does contain provisions for the rapid approval of medical devices that are "substantially equivalent" to preexisting devices. This clause has been used to gain rapid approval of many in vitro diagnostic tests using monoclonal antibodies that replace other antibody tests.

As might be expected, securing the approval of a new drug or biologic is usually a long and expensive process. The approval of an NDA by the FDA typically takes anywhere from six to eight years and costs tens of millions of dollars. However, in certain cases things can proceed much more quickly. It took only four years for the FDA to approve Eli Lilly's human insulin from the time it was first produced by genetically engineered *E. coli.*

According to Harry M. Meyer, Jr., of the FDA's Center for Drugs and Biologics, current legislation should be sufficient for FDA to regulate the products expected to emerge from biotechnology. "We feel that we can regulate the products of biotechnology on an individual basis under our existing authorities," he says. "For example, as part of the review of any new drug or biologic, the manufacturing process is carefully studied. We do a case-by-case review, and with that, nuances that influence safety and problems that could reduce the effectiveness can be identified and dealt with."

Nevertheless, according to Meyer, the FDA has been making changes in its staff and internal structure to interact more effectively with the biotechnology industry and academic researchers. In anticipation of future developments, the agency has established new research programs in areas related to biotechnology and has been bringing in additional scientific talent. It has also developed a series of "points to consider" documents for scientists working on specific products that will be submitted to the agency. Says Meyer, "Our regulation over the past several years of recombinant-produced human insulin, growth hormone, interferon, lymphokines, vaccines, and numerous products produced by hybridomas has been characterized, at least in my opinion, by what I see as our posture for the future—problem solving through joint efforts with industry and the use of scientific consensus to guide the direction of investigational efforts."

In addition to regulating human drugs and biologics for domestic use, the FDA oversees the export of these substances to foreign countries. In particular, the FDA interprets the Food, Drug and Cosmetic Act as forbidding the export for commercial purposes of new drugs or biologics that have not been approved in the United States. The FDA does allow the export under certain conditions of small amounts of new drugs or biologics for clinical testing. But once approved in a foreign country, a new drug cannot be exported for sale until approved in the United States.

This provision of the Food, Drug and Cosmetic Act, which has been retained in part to prevent companies from "dumping" ineffective drugs in foreign countries, has had an unfortunate effect on the U.S. biotechnology industry. A company wishing to sell an unapproved drug that has been approved in another country faces two unappealing alternatives. First, it can build production facilities in countries where the drug is already approved or where the law does not prohibit its export. This option is generally not available, however, for smaller biotechnology companies with limited resources. Second, a company can enter into joint agreements with foreign firms, supplying the production technology to the foreign partner in return for a share of the proceeds. The greatest damage caused by this second option may occur in the long run. "The prohibition compels the transfer of biotechnology to foreign countries," says Brian Cunningham of the biotechnology firm Genentech, "because the foreign partners must be given the ability to develop the bulk products themselves from the microorganisms that were genetically engineered in this country. This transfer of technology is not mitigated once the United States finally approves the new drug or biological product, because by the time U.S. approval is obtained, foreign production is already well under way. The foreign country continues to be the location from which world markets are supplied."

The FDA has joined with American manufacturers in backing legislation that would allow the export of unapproved new drugs or biologics, under strict conditions designed to prevent drug dumping, to countries where the substances have been approved, but past initiatives have been unsuccessful. "We believe that the governments of other nations are the proper authorities to assess their health needs, the diseases and health-related characteristics of their populations, the nature of their health care systems, and the availability of treatment alternatives," says Meyer. "In other words, we think they are the ones in the best position to make a benefit-to-cost decision about a drug or a biologic to be used in their country. However, for us to implement this philosophy would require a change in the law."

The Environmental Protection Agency

The Environmental Protection Agency (EPA) will regulate applications of genetic engineering primarily under the Federal Insecticide, Fungicide, and Rodenticide Act (FIFRA); the Toxic Substances Control Act (TSCA); and a handful of antipollution statutes directed at specific parts of the environment. Under current interpretations, these laws give the EPA authority over a wide range of chemical and biological products of biotechnology.

Like the Food, Drug and Cosmetic Act, FIFRA requires a manufacturer to demonstrate that the use of an insecticide, fungicide, or rodenticide—more generally, a pesticide—will not cause "unreasonable adverse effects" on human health or the environment. When satisfied that this criterion has been met, the EPA registers the product for use. In the past the EPA has taken the position that any microorganism used as a pesticide falls under FIFRA and has required premarketing registration for any such product. It will continue to apply this standard for microbial pesticides produced through genetic engineering, which it has defined very broadly to include not only recombinant DNA but cell fusion and a variety of other genetic techniques.

Under FIFRA, the EPA can also require that a manufacturer obtain an experimental use permit (EUP) before field testing a pesticide. In the past the EPA has generally not required an EUP for field testing on a small scale, which it has traditionally defined as less than 10 acres of land or 1 acre of water. However, because of the possibility of genetically engineered or nonindigenous microbial pesticides escaping from the bounds of a field test and multiplying, the EPA has decided that as an interim policy it will require companies to provide it with certain information at least 90 days before any field testing is begun. Such notification is not required for experiments in contained laboratories, growth chambers, greenhouses, or other facilities where there is no release of the microorganism into the environment.

The EPA has already begun to receive requests from companies for comments on limited field tests of genetically engineered microbial pesticides. Says John A. Moore of the EPA's Pesticides and Toxic Substances Division, "The theoretical is now. It is not off in the future." Like the FDA, the EPA plans to take a case-by-case approach to submissions by manufacturers, using the scientific capability within and outside the agency and referring significant issues and problems to a science advisory panel mandated by FIFRA.

Moore, like others, is "less than sanguine" about the uncertainties still surrounding the risk of releasing genetically engineered microorganisms into the environment. The only way to reduce these uncertainties, he says, is through research into the organisms and situations that are being considered. "What we have to do is to bring those people who are most knowledgeable on the subject to focus on the particular issues, and indeed give us the best guidance and judgment that they can, based on what we do know, then make what hopefully is the most appropriate judgment."

Although FIFRA is a strong law, its applicability is somewhat limited because it applies to genetically engineered microorganisms only if they are to be used as pesticides. Genetically engineered microorganisms used in the environment for other purposes, such as treating oil spills or toxic waste dumps, would not fall under FIFRA. To ensure adequate overview of these applications, the EPA plans to apply the Toxic Substances Control Act.

The Toxic Substances Control Act (TSCA) is essentially a gap-filling statute under which the EPA can regulate the production, distribution, use, and disposal of chemicals that it believes pose an "unreasonable risk" to human health or the environment. Unlike FIFRA or the Food, Drug and Cosmetic Act, TSCA does not require that a chemical be approved before it is marketed. Rather, it requires that a manufacturer submit a premanufacturing notice (PMN) before it begins to make a "new chemical substance," which is defined as a substance not on an inventory of existing substances. The EPA has 90 days to review the PMN and inform the manufacturer whether it will require additional data or testing on the effects of the substance on human health or the environment. If the EPA does nothing, the manufacturer is free to begin production.

The EPA has proposed to apply TSCA to genetically engineered microorganisms by defining "chemical substances" to include both DNA and the microorganisms that contain that DNA. (It plans to leave jurisdiction over plants and animals to the USDA and the Department of the Interior.) If this definition stands, it would give the EPA authority over a large segment of the biotechnology industry's products (although chemicals regulated by the FDA are exempted from TSCA). The agency could require a PMN not only for new genetically engineered microorganisms to be used in the environment but also for those to be used in fermentation processes. However, the EPA has not yet taken a final position on its definition of "new." One possibility it has proposed is to define chemical substances produced by recombinant organisms as new to prevent those substances from slipping through cracks in the regulatory framework.

The Toxic Substances Control Act exempts "small quantities" of new chemicals from overview if used solely for research and development. But because of the possibility that genetically engineered microorganisms might transgress the bounds of a field test and multiply, the EPA is considering defining any quantity of such organisms used in field tests as not small. Certain kinds of research involving genetically engineered microorganisms would therefore always require a PMN. As with other aspects of its proposed regulation of biotechnology, the EPA has sought public comment on this issue.

The categorization of DNA and genetically engineered microorganisms as "new chemical substances," although generally supported in Congress and in the biotechnology industry, is controversial and may be challenged in court. TSCA is also not as strong a statute as FIFRA or the Food, Drug and Cosmetic Act because the burden of proof is on the EPA rather than the manufacturer to prove "unreasonable risk." Largely for these reasons, Congress has held hearings to determine whether additional legislation in this area is necessary to ensure adequate regulation of biotechnology.

But TSCA also has "a large number of strengths that should not be overlooked," says Geoffrey M. Karny, a Washington, D.C., attorney and former senior analyst at the Office of Technology Assessment. "First of all, it is designed as an information-gathering statute. It is a middle-of-the-road approach between no regulation or voluntary regulation and some kind of stringent premarket approval mechanism. It is designed to be flexible, to deal with substances on a case-by-case basis, and to accommodate change in safety data. Finally, it involves a balancing, because the operative language is 'unreasonable risk.' "

Like the FDA, the EPA has been building up its staff in anticipation of the flood of products expected from biotechnology. Some still question, however, whether the agency will have the manpower to effectively regulate the wide variety of substances over which it has claimed jurisdiction. The question has also been raised of whether government agencies will have access to the information and technologies needed to evaluate the claims of manufacturers working in biotechnology.

The U.S. Department of Agriculture

The U.S. Department of Agriculture (USDA), under a variety of statutes, regulates the importation and interstate shipment of broad categories of plants, animals, and agricultural microorganisms. It has stated that it plans to treat products produced through biotechnology in the same way that it treats products produced through conventional means. It does not expect to encounter any problems unique to applications of biotechnology, although it does plan to continually reevaluate its position as the state of the art evolves.

The department also has authority over animal biologics, which in the past has created some overlap and conflict with the FDA. The two agencies have developed a "memorandum of understanding" in an attempt to resolve jurisdictional disputes. The USDA also claims authority over microorganisms that are plant pests or pathogens, bringing it into potential conflict with the EPA.

The USDA has the most experience of any federal agency regarding the introduction and monitoring of novel plants, animals, and microorganisms in the environment. According to Orville G. Bentley, the USDA's Assistant Secretary for Science and Education, it also has an extensive research network that it can call on in evaluating agricultural products developed through biotechnology. "This institutional expertise and capability," says Bentley, "will serve as a powerful source in the regulatory process and in averting any particular problems that might come as a result of the application of biotechnology."

Other Agencies and Legislation

A number of other federal agencies also have either a direct or indirect influence over the development or use of biotechnology. The

National Institute of Occupational Safety and Health, the Centers for Disease Control, and the Occupational Safety and Health Administration oversee the health and safety of people who work with recombinant molecules or organisms in commercial settings. The Department of Defense has been sponsoring genetic engineering research, all of it unclassified, on various medical and materials problems of interest to the military. The Agency for International Development sponsors research on how genetic engineering might be applied to solve problems that occur in less developed nations. The Department of Commerce maintains export controls over biological research materials in an attempt to keep certain items from reaching the Eastern bloc. Other federal agencies with various degrees of sway over genetic engineering and biotechnology include the National Science Foundation, the Patent Office, the Department of Energy, the Department of the Interior, the National Bureau of Standards, and the Department of State.

An important piece of federal legislation that has already left its mark on the development of biotechnology is the National Environmental Protection Act (NEPA), which requires federal agencies to prepare environmental impact statements for major actions that significantly affect the environment. In the past this act has led to a substantial amount of litigation, some of it designed only to harass or delay a proposed project. According to Karny, it is the responsibility of the courts to see that NEPA is used for its intended purposes rather than for purely obstructionist reasons. "I think judges have an obligation to throw out frivolous lawsuits in no uncertain terms," comments Karny. "Hopefully, they will be encouraged to do so if they have confidence in the existing regulatory system, especially if they see that it operates to accomplish the goals of the National Environmental Protection Act without all of the formalities."

Finally, there is a large body of state tort law, which allows for private lawsuits for damages caused by a civil wrong. Because of the potentially large damage amounts involved, private lawsuits can provide a strong incentive for safety-conscious conduct by companies.

The Cabinet Council Working Group on Biotechnology

In response to perceived gaps, conflicts, and inefficiencies in the regulation of biotechnology, the Cabinet Council on Natural Resources and the Environment formed the Cabinet Council Working Group on Biotechnology in April 1984 under the leadership of the White House Office of Science and Technology Policy. The working group's mandate was to review current regulations and policies affecting biotechnology, determine if additional regulation was necessary, and develop recommendations for administrative and legislative actions to resolve identified problems. In its own words, the working group sought to point the way toward "a coordinated and sensible regulatory review process that will minimize the uncertainties and inefficiencies that can stifle innovation and impair the effectiveness of U.S. industry."

In December 1984 the working group published a detailed description of the federal laws and proposals affecting biotechnology and its proposed framework for the future regulation of biotechnology. The proposal called for a two-tiered science review system based on the expertise and flexibility demonstrated by the RAC. Scientific advisory committees modeled on the RAC would be established in each of the five agencies that have significant jurisdiction over biotechnology: the Food and Drug Administration, the Environmental Protection Agency, the U.S. Department of Agriculture, the National Science Foundation, and the National Institutes of Health. In several cases, committees already in existence at these agencies would continue to serve slightly modified roles. These committees, composed of recognized experts in disciplines related to biotechnology, would do the detailed case-by-case review of individual submissions, observing the needs of their agencies with regard to time constraints and confidentiality. The NIH and NSF would concentrate on scientific research, while the FDA, EPA, and USDA would focus on the commercial products emerging from biotechnology.

In addition, each of these committees would provide information to an interagency coordinating committee on biotechnology—sometimes dubbed the super-RAC. This committee would consist of members from the agencies' committees along with other scientists and nonscientists as appropriate. The coordinating committee would review the summary reports of the individual committees and would have the option of recommending that a specific application be reviewed by another agency. It could also conduct analyses of broad scientific or social issues with an eye toward developing generic guidelines applicable across the entire field. The committee would be subject to periodic review to determine if it should continue to exist.

The working group envisioned that the coordinating committee would provide direction to the scientific research underlying the regulation of biotechnology. "A central core of scientific expertise for all the agencies, we believe, would promote consistent assessments of risk," says Healy, the working group's chairman. "That is a key to final analysis and streamlined regulatory decision making." At the same time, the committee, like the RAC, would remain purely advisory in nature. "It provides the necessary input to the regulatory process, but it does not make the regulatory decisions. [That] is the responsibility, under law, of the agencies."

While many people have lauded the recommendations of the working group, others have found fault. Specifically, the establishment of an interagency coordinating committee to oversee committees within the agencies has been viewed as posing the risk of "redundant review and the interpolation of additional layers of bureaucracy," according to Irving S. Johnson of Eli Lilly and Company. "We feel that instituting this two-tiered approach may have a serious impact on and inhibit the development of biotechnology and its products, and perhaps compromise our competitive position in the international arena."

Johnson and others also question the need for a proliferation of scientific advisory committees in the regulatory system. In their view, the RAC is well equipped to continue to provide the exemplary guidance it has provided in the past. If anything, argues Johnson, the functions of the RAC should be expanded, making it in essence the super-RAC proposed by the working group. "I would strongly urge that the RAC or its equivalent be allowed to continue as a single oversight group, and that the remainder of the system be adjusted to accommodate it," he says.

According to Johnson, the RAC has a number of resources that it could call on if it were to assume such a role. Several institutes within NIH could provide technical assistance in areas related to the commercialization of biotechnology. As in the past, the RAC could also enlist outside consultants and form working groups to deal with topics of special interest. Finally, the workload of the RAC will not necessarily increase, because many of the major concerns associated with biotechnology have already been resolved. "Most of the serious generic issues have been or are in the process of being addressed by the RAC," says Johnson. "After deliberate release of microorganisms and the concept of gene therapy, I am not sure what the next major issue is going to be."

The Role of Congress and the Public in Biotechnology

The Cabinet Council Working Group on Biotechnology concluded that no new legislation was needed to give federal agencies adequate regulatory authority over the anticipated products of biotechnology. Many legislators and industry leaders have reached similar conclusions. "I continue to believe that no new legislation is needed at this time," says Senator Albert Gore, Jr. "The various agencies all seem to feel that they have adequate statutory authority to do the job. As long as the current evaluation continues on a serious course, I think the necessary oversight can occur with minimal legislative adjustments."

Despite the prevalence of this opinion, interest in biotechnology has recently swelled in Congress, and it can be expected to remain strong, according to Robert P. Nicholas, former staff director of the Subcommittee on Investigations and Oversight under the House Committee on Science and Technology. Legislation may be forthcoming in such areas as the export of unapproved drugs and biologics, human gene therapy, patent laws, and environmental risk assessment. In addition, the hearings held over the past few years on a wide range of subjects related to biotechnology, from environmental release to university-industry relations, testify to the widespread feeling in Congress that developments in biotechnology should be closely monitored.

In part, this continuing governmental oversight of biotechnology reflects an aspect of the field that has been apparent since Asilomar. Because of biotechnology's close association with life's most fundamental processes, public concern over developments in biotechnology must always be taken into account, whether by the scientific community, by industry, or by the government. "Many, if not most, of the questions that regulatory agencies are going to have to deal with when they assess the risks and benefits of these new technologies have strong public policy components," says Thomas O. McGarity of the University of Texas School of Law at Austin. "In the end, whether or not these new biotechnologies really get off the ground in this country is going to depend upon whether we can erect a regulatory regime that can secure public trust."

"Biotechnology is at a turning point," explains Nicholas. "The same questions that were being asked previously are again being asked, and they have become more important, since the end result of most of the [ongoing] industrial activity will be some release of a product. Unless these questions are sensitively addressed from here on out, there is a real risk that the consensus that has underpinned the development and financial support of biotechnology will recede."

According to William Ruckelshaus, a critical element in building the public trust necessary for biotechnology to prosper is public education. He maintains that the public must be informed "fairly, honestly, and straightforwardly" about both the potential benefits and the potential risks of biotechnology. The risks of a new technology inevitably become known at some point, and if the public has not been adequately informed about these risks, they may turn to any of a number of tactics available to slow down or halt the progress of a new field.

As a first step, public education could focus on the wide range of applications of biotechnology, differentiating the issues that are involved in each. "We must really talk about not biotechnology in the singular but biotechnologies," says Zsolt Harsanyi of E. F. Hutton and Company. "We have a variety of technologies that are quite disparate. In one case, you are talking about the insertion of a gene into a human being; in another case, you are talking about using immobilized enzymes to produce high-fructose corn syrup; in other cases, you are talking about microbes that might be released into the environment. . . . You have to get down to some very specific points and say what it is about any particular use that is going to be unique."

The best way to go about educating the public, according to Ruckelshaus, would be through an "elaborate, comprehensive, and sophisticated communication plan." Such a plan should recognize that different audiences require different messages. It could focus first on those who will most directly affect biotechnology—Congress, the regulatory agencies, industry, the press, environmentalists, academics. It could also take advantage of specific events—an important technical development, a particular experiment, regulatory approval of a product—to further public understanding of the field. Taken together, such efforts could begin to close the gap that has traditionally existed in the United States between scientific developments and public understanding. Concludes Ruckelshaus, "We need to do a much better job, not just in this area but across the board, as we try to grapple with the complexity involved in public participation in decisions of enormous scientific uncertainty."

Additional Readings

Thomas O. McGarity. 1985. "Regulating Biotechnology." *Issues in Science and Technology* 1(Spring):40-56.

Office of Science and Technology Policy. 1984. "Proposal for a Coordinated Framework for Regulation of Biotechnology." *Federal Register* 49 (December 31):50856-50907.

Sandra Panem and Hans Weill, eds. In press. *Biotechnology: Implications for Public Policy*. Washington, D.C.: Brookings Institution.

Joseph Perpich, ed. In press. *Biotechnology in Society: Private Initiatives and Public Oversight*. New York: Pergamon Press.

7

The New Biotechnology Firms

TWO DISTINCT TYPES OF FIRMS are pursuing the commercial applications of genetic engineering in the United States: the small
start-up companies founded primarily since 1976 to capitalize
specifically on genetic engineering research, and the established
multiproduct companies in such sectors as pharmaceuticals, chemicals,
agriculture, energy, and food processing that have invested in the field.
The interplay between the two and the complementary efforts of each
have done much to give the United States its current lead in
biotechnology.

Between these two types of firms, considerable amounts of money
have been invested in biotechnology. Since 1976, several billion dollars
have been funneled into the start-up biotechnology firms, of which
there are now more than 200. And since about 1981 many established
firms have set up major in-house biotechnology programs. Although it
is not always easy to characterize a specific industrial undertaking as
"biotechnology" (which makes the "biotechnology industry" similarly
hard to define), it is undeniable that substantial sums have been
devoted to commercializing the techniques of genetic engineering.

For the past several years, predictions of a shakeout among the

This chapter includes material from the presentation by Hubert J. P. Schoemaker
at the symposium.

start-up firms have dogged the biotechnology industry, contributing to a wariness among investors after a surge of enthusiasm for the new firms in the early 1980s. Yet remarkably few of these firms have gone bankrupt or been acquired by other companies. Nevertheless, most observers agree that mergers, acquisitions—even failures—can still be expected as new challenges arise within the industry. For one thing, the new firms will face increasingly stiff competition from the established firms. Also, many of the small firms will eventually reach the stage where to survive they will have to engage in production and marketing as well as research and development. Their degree of success in making this transition will have an important effect on the future of the industry.

Characteristics of the New Firms

Historically, small firms have often established the prominence of the United States in emerging advanced technologies, and biotechnology seems to be no exception. Within just a few years of the first experiments with recombinant DNA, small firms were being established, often by distinguished academic scientists, to commercialize the new techniques. Since then, the expansion in the number of new biotechnology firms has far exceeded the expectations of the technologies' founders.

The start-up firms have been working on projects that span the spectrum of biotechnology's potential applications. The most popular application to date has been the development of monoclonal antibodies for use in research, chemical separation and purification, diagnostic tests, and the treatment of disease. Many new biotechnology firms are also working to develop human and animal pharmaceuticals, which tend to have much higher costs associated with their development, regulatory approval, and production. Relatively fewer firms are working on such applications as commodity chemicals production or waste management, generally because much more research is needed to demonstrate the commercial feasibility of these pursuits.

To finance their research and development efforts, the new biotechnology firms have called on a wide array of funding mechanisms. Among the most important of these have been investments from venture capital firms and from established companies interested in biotechnology. The investments from the latter have generally taken two forms: equity investments and joint ventures. Equity investments, in which established companies buy portions of new biotechnology firms, have enabled the former to keep abreast of developments in the field, perhaps to gauge the best time to enter the field themselves. Joint ventures, in contrast, usually involve a more active combination of R&D contracts and product licensing agreements. Under the terms of these agreements, an established firm often handles the regulatory approval, manufacturing, and marketing of a product after the small firm has done the initial development. The small firm receives royalties from the sale of the product and usually retains the patent on the product.

In recent years, some new biotechnology firms have tried to lessen their reliance on R&D contracts and licensing agreements with large U.S. firms to retain more control over the uses and profits of their products. One way for them to do this has been to establish joint ventures with foreign companies. In these cases, the start-up firms often retain the rights to sell their products within the United States while selling the overseas sales rights to their foreign partners. In turn, the start-up firms supply either the products or the technology to make the products to the foreign companies. (As explained in Chapter 6, pharmaceuticals not approved in the United States can generally not be exported to another country for sale.) Many observers have questioned the wisdom of this transfer of technology, claiming that in the long run the spread of know-how generated in the United States to other countries will enhance the competitiveness of foreign firms. But most of the new biotechnology companies have deemed the short-term benefits of such an arrangement to be more important than the long-term disadvantages.

Another major source of funding for the new biotechnology firms has been the stock market. In the early 1980s several start-up biotechnology firms set Wall Street records when they first went public. Genentech's stock underwent the most rapid price increase in the market's history, climbing from $35 to $89 per share in its first 20 minutes of trading. A few months later, Cetus raised the largest amount of money that has ever been made with an initial public offering—$110 million. A couple years later the glow had faded from biotechnology stocks and they were trading for much less than their previous highs. But recently prices have rebounded, and the stock market remains a promising source of revenues for biotechnology firms. In fact, several firms have returned to the market two and three times to finance production scale-ups and clinical trials of their products.

A source of financing that has rivaled the stock market in size is a type of investment known as an R&D limited partnership. This allows individuals or organizations to invest in a company's research and development and to write off that money as expenses. The investors become limited partners and are entitled to receive royalty payments from future sales of products. Part of or all these royalties are in turn taxed as capital gains, offering an added attraction to this kind of investment.

The new biotechnology firms also have a number of other sources of capital, including interest from funds previously raised, short-term loans, industrial revenue bonds, and equipment leasing. Through these and other funding mechanisms, the new biotechnology firms have generally been able to bring in enough revenue to remain viable, even though many of them have not yet generated actual products for the marketplace.

Portrait of a Successful Firm

One new biotechnology firm that has generated products for the marketplace through its work with monoclonal antibodies is Centocor, Inc., which was founded in 1979. According to the company's president, Hubert J. P. Schoemaker, Centocor's approach is based on a careful analysis of several key features of the health care business. First, it is very expensive to produce and market pharmaceutical products, but very few products are needed for the firm to be successful. Second, most health care companies have a relatively narrow product focus, since this optimizes the distribution of products and reduces risk. Third and most important, the health care industry worldwide currently has excess capacity in its manufacturing, distribution, and sales networks. As a result, "these companies are looking for products to feed into their investments," says Schoemaker. "There is a product shortage. This leads to a fairly aggressive acquisition strategy and to an aggressive licensing-in strategy. . . . Centocor, as a company, and, I believe, similar companies, were formed to capitalize on these features of the industry. Centocor bridges the gap between new innovations and the already existing product distribution networks."

To find products to feed into these networks, Centocor's in-house technical groups keep close tabs on the research being done in universities and in public and private research institutes. When the company uncovers work with commercial promise, it seeks to establish collaborative agreements with the investigators or research institutes that have done the work. Currently the company has initiated agreements with about 30 universities around the world to gather the results of research.

Once Centocor has obtained the rights to a research development,

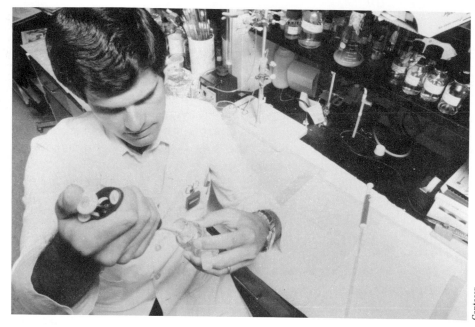

Centocor

A researcher at Centocor prepares a monoclonal antibody assay. Many of Centocor's products are based on research done in universities that the company then develops for the marketplace.

the project is brought into the company, where Centocor's own staff develops the product, performs any clinical tests needed to have it approved in world markets, establishes the market for the product, and introduces it for commercial sale. According to Schoemaker, the company can usually ready a monoclonal antibody blood test for sale within four to six years of the antibody's development. "In our first six products the critical raw material, the antibody, was developed within a university, licensed, and brought into the company, the product was developed, and we now are delighted to pay royalties to these institutions," says Schoemaker.

After Centocor has developed and introduced a product, the company often licenses it to pharmaceutical companies to supply a raw material like an antibody or sell an end product like a blood test. In this way, Centocor can take advantage of the unused capacity of existing production and distribution systems without having to overextend its own. The company has tried to stay away from exclusive arrangements with its partners, preferring to rely on territorial rights or a particular product format. "This goes a little against the culture of the health care industry, which likes worldwide and exclusive rights, but we believe we lose some control over our own destiny doing that," explains Schoemaker. Centocor also gives its partners the option of using the raw materials it provides in the partners' own product configurations. That extends the lifecycle of a product that Centocor has spent considerable effort developing.

Centocor's initial products were in vitro monoclonal antibody tests for hepatitis B, ovarian cancer, gastrointestinal tract cancer, and breast cancer. Building on those successes, it has been developing similar tests for cancers of the colon, liver, and lung. Another product line involves monoclonal antibodies to be used inside the body for imaging both the location and extent of diseases such as cancer and atherosclerosis. A third product line involves the use of monoclonal antibodies for therapy. This wide range of pursuits has left Centocor with a problem that is unusual in the biotechnology industry. Says Schoemaker, "We have too many products for the size that we are. Most of these products are quite innovative and require significant market development to get them introduced worldwide."

Half of Centocor's $3.2 million in sales during 1984 were in Japan, with the other half split between Europe and the United States. In addition, the company relies for its income on research contracts ($7.6 million in 1984) and on the interest ($2 million) from the remaining $18 million of $21 million raised in a public stock offering in 1982. With these three sources of income, the company earned its first profits in 1984.

According to Schoemaker, Centocor intends to continue to act as a product development company that links academic research to health care distribution networks. "We believe that in biotechnology the innovations will still be made to a large extent within the academic realm. We have developed an organization that can capitalize on these developments. Centocor has built a company that can very quickly commercialize this technology and deliver it to the health care industry through the existing distribution channels. This strategy appears at this time to be successful."

Challenges to the New Firms

The young firms that are striving to make their mark in biotechnology will encounter a number of difficulties as the industry enters its second full decade. For one, the competition posed by established firms can be expected to heighten. Until about 1981 the large established firms generally stood on the sidelines in biotechnology, content to monitor developments in the field through various kinds of arrangements with the new biotechnology firms and academic research centers. But during the last few years they have moved into biotechnology in force. In 1984 both Monsanto and du Pont opened major life science research facilities representing a combined investment of more than $200 million. Such resources dwarf the amounts available to any new biotechnology firm. However, the established firms have continued to invest in the fledgling companies. Some new firms have also been protecting themselves by moving toward more limited product niches that would not interest larger firms.

Perhaps a more serious problem lies in the *nature* of the biotechnology industry. To compete with the established companies and with other new biotechnology companies, the start-up firms will eventually have to become profitable through the sale of their products. In the early stages of the industry, firms have taken different approaches to this basic requirement. Some, like Centocor, have licensed part of the production and marketing of their products to established firms. Others are licensing all their technology to established companies in exchange for royalties. Still others, like Genentech and Cetus, are attempting to become fully integrated manufacturers and distributors.

Fully integrated companies will continue to rely on the sources of revenue that have seen them through their formative years as they conduct clinical trials, set up production facilities, and organize marketing systems. In addition, as discussed in Chapter 10, the actions of the federal government could significantly influence the future courses of all new biotechnology firms.

Additional Readings

"Biotech Comes of Age." 1984. *Business Week* (January 23):84-94.

Peter Hall. 1984. "The Business of Biotechnology." *Financial World* (March 21-April 3):8-14.

Ralph W. F. Hardy and David J. Glass. 1985. "Our Investment: What Is at Stake?" *Issues in Science and Technology* 1(Spring):69-82.

Arthur Klausner. 1985. "Corporate Strategies: And Then There Were Two." *Bio/Technology* 3(July):605-612.

U.S. Department of Commerce, International Trade Administration. 1984. *High Technology Industries: Profiles and Outlooks—Biotechnology*. Washington, D.C.: U.S. Government Printing Office.

8

Patents and Trade Secrets in Biotechnology

A PRIME CONCERN AMONG FIRMS throughout the biotechnology industry is the degree of protection they can obtain over the products and processes they develop. In the United States several means exist for securing such protection. An individual or organization can receive a patent for a product or process, which gives the holder the right to exclude others from making, using, or selling the invention for 17 years. Patents are also available for asexually reproducing plants, and certificates (rather than patents) are granted for sexually reproducing plants. An individual or organization may also elect to keep an invention secret, with recourse to legal proceedings if others acquire the secret through improper means.

In 1980 the Supreme Court ruled by a vote of five to four in *Diamond v. Chakrabarty* that a particular genetically engineered microorganism could be patented. (The organism in question had been genetically engineered not through recombinant DNA techniques but through the transfer of naturally occurring plasmids.) This landmark decision led to a surge of patent applications and approvals in the area of biotechnology. However, the decision also left a number of important questions unresolved. For instance, it did not explicitly determine if

This chapter includes material from the presentation by Roman Saliwanchik at the symposium.

higher organisms, including plants and animals, could be patented under the same provisions as microorganisms, although it does seem likely. Currently, patents for asexually reproduced plants are granted under a different system, and it is not clear which system will apply to genetically engineered plants. The court also did not specify the extent or nature of human intervention required to make an organism patentable. So far the Patent and Trademark Office has taken a very conservative approach in granting patents for genetically engineered organisms.

Nevertheless, the *Chakrabarty* decision has sparked new interest in using the patent system to protect intellectual property resulting from biotechnology. As the ramifications of the decision continue to spread, and as new court cases and legislative initiatives arise, many of the remaining uncertainties and problems in protecting innovations in biotechnology will diminish. And if, in due course, the applicability of the patent system to biotechnology is strengthened, the result should be a stronger, more productive industry.

The Patent System and Biotechnology

The rationale behind a patent system is simple; the implementation of that rationale can be very complex. Essentially, a patent grants its owner a monopoly over the use of an invention for a given period of time. In this way, patent systems, which have existed in various forms since antiquity, reward the effort and risk that are required for innovation. In return for a patent, an inventor must disclose the nature of the invention so that knowledge of it will not die with the inventor and so that society can make free use of it once the patent has expired.

To be patentable under U.S. law, an invention must meet several criteria. First, it must be useful, which the courts have defined broadly enough to not pose great problems to biotechnology. Second, it must be new, so that a patent is not granted for something that already belongs to the public. Third, it must be nonobvious, so that a person skilled in the field cannot take something already in the public domain, add a slight twist to it, and receive a patent for the result.

Materials and knowledge already in the public domain at the time the patent application is filed are referred to as the "prior art." This is a critical concept in determining both the novelty and nonobviousness of an invention. In deciding if a claimed invention meets these criteria, a patent examiner must evaluate it from the perspective of the prior art and determine if it departs sufficiently from that domain. According to Roman Saliwanchik, a patent attorney in Richland, Michigan, who specializes in biotechnology, it is often particularly difficult in biotechnology to keep from overlapping the prior art. "There have been a tremendous number of publications in the genetic engineering area," he says. "All of these publications can be used as prior art against the inventor if published prior to the filing of a patent application."

The novelty requirement does not mean that things already existing in nature cannot be patented, according to Saliwanchik. "The only requirement is that the material be novel," he says. "When you purify a material and it has never existed in that form before, it is novel." This provision has been used to patent a wide variety of chemical substances in the past, including vitamins, hormones, and pure cultures of naturally occurring microorganisms, and it applies equally well to the products of biotechnology.

A patent application must also contain enough information to enable a person who is skilled in the field to make and use the claimed invention without undue experimentation. This provision, which is known as the enablement requirement, clarifies the nature of the invention so that others will not infringe on it and ensures that the public will eventually receive full benefit from the patented knowledge.

The enablement requirement raises certain difficulties in biotechnology. It is generally impossible to explain in writing how a microorganism performs a given task in such a way that others will be able to recreate that microorganism. Some courts have therefore ruled that to meet the enablement requirement the patent applicant must usually deposit a culture of the microorganism in a public repository. These repositories are in turn required to provide samples of the culture on request to members of the public after a patent has been issued.

In biotechnology, this means that a competitor has access to the very heart of an industrial process—the microorganism that performs a biological function. An unscrupulous competitor could acquire that microorganism and exploit it secretly, or change it slightly and put it to work performing the same task. Such possibilities have made some entrepreneurs in biotechnology selective in seeking patent protection.

A patent application generally concludes with a series of claims that define the invention with varying degrees of specificity. These claims establish differing boundaries for what the patent covers. In its examination of the application, the Patent and Trademark Office decides which claims it will allow, and those establish the scope of the patent.

This measure of a patent's applicability can be crucial in biotechnology. "Just getting a patent may not be enough," explains Saliwanchik. "If you don't get the right scope on your claims, you can have a weak

U.S. Department of Agriculture

A researcher withdraws a cell line sample from a freezing device at the American Culture Type Collection in Rockville, Maryland. Under U.S. patent law, samples of patented microorganisms must generally be provided to independent repositories, which then furnish the cell lines to members of the public on request.

patent." For instance, if the scope is not properly defined in a patent, a competitor may be able to use and possibly patent a protein with a different combination of amino acids that has essentially the same biological function. "I think these kinds of things are going to be bounced around a bit," he says. "It is what is called 'scoping.' We have gone through a long history of scoping in the chemical arts, and we pretty well know our limits. But in the biological arts there is an element of unpredictability that sometimes makes scoping more difficult."

Despite the remaining uncertainties, Saliwanchik believes that the patent system is well equipped to deal with the onslaught of applications coming from biotechnology. For one thing, it can draw on its extensive past experience in issuing patents on metabolites, proteins, vaccines, and even some processes using naturally occurring microorganisms. As an example, he cites the patenting of antibiotics. "The antibiotic industry itself would probably not have prospered to the extent that it has but for a strong patent system. We have to have that same strong patent system for the genetic engineering industry."

Already the Patent and Trademark Office has issued patents for a wide range of products resulting from biotechnology, including altered genes, DNA probes, vectors, and microorganisms. As for processes, it has issued patents covering such diverse procedures as the enhanced expression of a protein, the alteration of gene components, the preparation of vectors, the synthesis of proteins, the production of hybrid bacteria, and the purification of DNA sequences. "We are getting a broad range of patents issued to cover genetic engineering processes," Saliwanchik says.

Monoclonal antibodies and the hybridoma cell lines that produce them have also been patented. "This subject matter can be covered very well by patents," notes Saliwanchik. "You can cover the antibodies themselves by having the proper characteristics defined in the specification of the application for the patent. You can also claim the cell line. Certain procedures must be followed in doing this, but the procedures are known to patent attorneys."

Thus, Saliwanchik contends that the patent system has been responding well to biotechnology. "Some people have expressed doubts as to whether or not the system is viable enough to cover genetic engineering technology. I think it is."

Trade Secrecy and Biotechnology

If an individual or company decides not to pursue patent protection for a product or process, it may rely on trade secrecy instead. Trade secrets can theoretically be held indefinitely, they don't have to meet the criteria for patentability, and they may be less expensive to maintain and enforce than patents. They are enforced through state laws, with the holder of a trade secret able to obtain either an injunction or monetary damages against an unauthorized user of the secret when it has been acquired through improper means.

An individual or organization must consider a number of factors in deciding whether to patent an invention or to keep it secret. Because biotechnology is developing so rapidly, a given invention may be outdated before a patent can be issued (typically the process takes two years or longer). Companies are also more likely to patent a pioneering invention, one with the potential to change the direction of the field, than they are a less dramatic development or an improvement in the state of the art.

There are also a number of drawbacks associated with trade secrets that may dissuade a company from pursuing that option. Some states are less protective of the results of research, as opposed to trade secrets of obvious commercial value. A trade secret can also be inferred through reverse engineering. If this reveals the secret or if it is independently discovered, the new discoverer may patent the secret and prohibit the original discoverer from using it. Furthermore, it is often difficult to prove in court that a competitor has improperly acquired a trade secret.

Trade secrecy also runs counter to widely held tenets within the scientific community. It requires, for instance, that researchers not publish their findings in the open literature or discuss them at scientific conferences. "Once you file a patent application, you can publish on what you have done," says Saliwanchik. "Once you start thinking about trade secrets, you can forget about publishing, because that would destroy the trade secret."

In biotechnology, trade secrecy may also conflict with requirements to release information in public forums. To build public trust in the new techniques, it may be necessary to release to regulatory or public groups information that a company would rather keep secret.

Developments in the Patent System

The patent system is continually evolving and can be expected to adapt both to developments in biotechnology and to more general influences. With regard to the former, biotechnology is forcing a reevaluation of the applicability of the enablement requirement to living organisms. One possible modification of this requirement would be to deposit the original organism and file an enabling description of how it has been genetically altered. Another option may be to restrict the use of microorganisms by third parties in a way that prevents patent infringement without limiting the public's access to the patented information. "I think this issue will have to be resolved," says Saliwanchik. "I would hope that it is resolved in a way that strengthens the patent system."

Another prominent issue involves patents on processes. A foreign competitor can use a process that is patented in the United States and then legally ship products made through that process into the United

States for sale. Legislation has been introduced that would define this practice as patent infringement, but it has not yet been passed. Because the manufacture of a product with genetically engineered organisms is a process, such legislation is strongly supported by American biotechnology firms.

Meanwhile, changes are occurring *within* the Patent and Trademark Office in response to biotechnology. According to Saliwanchik, that office has been hiring new staff and training its examiners to deal with the flood of patent applications in biotechnology. But he warns that companies must realize that even with an expanded staff the granting of a patent takes time. "People think that if the Patent Office doesn't function rapidly to get a patent out in two years, then the system is defective. That is not true."

The Patent and Trademark Office will also have to work through several uncertainties in its treatment of biotechnology, such as its assessment of the prior art and how to resolve conflicting claims. These problems will undoubtedly diminish as examiners and patent attorneys become more knowledgeable about the field and as precedents are established. But "it takes time," says Saliwanchik, "upwards of ten years or so in conflict situations."

Finally, the cost of seeking a patent has to be kept within bounds so that individuals and organizations will take advantage of the system. "The system has to be affordable for everyone," says Saliwanchik. "There is a tendency for prices to go up, and I think we have to try to keep those prices down."

Despite these several deficiencies in the patent system, Saliwanchik does not foresee any great difficulties in applying for and receiving patents for products and processes in biotechnology. "I think the patent system in the United States is in pretty good shape," he says. "There may be some problems along the way, but I think we have reason to believe that the protection will be strong."

Additional Readings

Reid G. Adler. 1984. "Biotechnology as an Intellectual Property." *Science* 224 (April 27):357-363.

D. W. Plant, N. Reimers, and N. D. Zinder, eds. 1982. *Patenting of Life Forms.* Banbury Report 10. Cold Spring Harbor, N.Y.: Cold Spring Harbor Laboratory.

Roman Saliwanchik. 1982. *Legal Protection for Microbiological and Genetic Engineering Inventions.* Reading, Mass.: Addison-Wesley.

9

University-Industry Relations

THE BIRTHPLACE OF BIOTECHNOLOGY, and the source of much of its continuing inspiration, is academia. Researchers in university biomedical laboratories, funded largely by the federal government, developed almost all the basic techniques that have given rise to biotechnology. In turn, these researchers were among the first to recognize the commercial potential of the new techniques, and many of them were among the founders and first employees of the new biotechnology companies.

All the firms involved in biotechnology remain vitally interested in the research being conducted in universities and other institutions. These companies view the universities as sources of new ideas and innovative techniques, as suppliers of trained employees and talented consultants, and even as customers for research equipment and other products. The extremely rapid pace of research in molecular biology puts a premium on staying at the forefront of the field. Companies realize that one of the best ways to do this is through exposure of their scientists to people working on basic research problems. Furthermore, such exposure can provide these companies with a window on the

This chapter includes material from the presentation by David M. Kipnis at the symposium.

technology while not necessarily requiring large investments in facilities and personnel.

Less well known than the benefits to individual companies are the benefits to universities that derive from university-industry contacts. Exposure to problems and opportunities in industry creates new challenges for academic science and engineering, places undergraduate and graduate education in new perspectives, increases scientific communication and cooperation, and ties educational and research programs of universities more closely to national and regional needs. Also, in a time of stagnating or declining federal outlays for scientific research, universities look to alliances with industry as a way to diversify and supplement their sources of funding.

Industry is currently providing between $200 million and $300 million to universities for research and development of all kinds, and most major universities have set up special offices to seek and administer these funds. However, this represents only about 3 to 4 percent of the universities' total funding for research and development (although uncategorized forms of support may raise that figure to 6 to 7 percent). All observers agree that industrial support of research in universities cannot and should not substantially replace federal support of university research. But, by the same token, industrial support of this research can have an effect above and beyond its relative size, especially in its encouragement of the complex process of technology transfer between basic research and commercial application.

Molecular biology is far from the first discipline in which university-industry relations have become a significant issue. Such alliances have been common for many years in agriculture, chemistry, physics, electrical engineering, medicine, and the defense-related industries. Thus, although the problems that arise in biotechnology may differ in scale or in subject matter, they do not differ in kind from problems that have arisen and have been successfully resolved in the past.

Types of Agreements and Potential Problems

The types of alliances that have sprung up between the biotechnology industry and universities are as varied as the institutions and individuals that engage in them. They exist on a wide range of scales—from the single investigator to consortia of companies and universities—and they tend to reflect the characteristics and concerns of the entities involved.

On an individual level, consulting arrangements and extension services are a common form of interaction between universities and industry. Most universities allow their faculty members to consult for industry and have established guidelines to govern these activities. On a larger scale, universities and industry may set up industrial associates programs, in which a company pays a fee to a university in exchange for the right to participate in part of the university's research activities. This may entail attendance at seminars or classes, interactions with faculty and students, or the preview of publications.

Universities may also sign a contract with a company to conduct research that the company will then use. Such contracts may affect a single researcher or an entire university department; they may call for basic, open-ended research or for more applied research directed toward a specific goal. Industries and universities can go even further and set up research foundations, university consortia, or industry cooperatives, all designed to establish more long-term or independent research institutions. Such institutions are often actively sought by state and local governments eager to encourage the establishment of high-technology industry within their areas. Finally, some universities have formed innovative, and often controversial, private corporations affiliated with the universities so that faculty can remain on campus while working on the commercial applications of their research.

The specific concerns associated with university-industry relations depend largely on the type of alliance under consideration. But there are several generic categories of problems that have arisen in the past—problems that should be carefully thought out before any type of alliance is formed.

The most fundamental of these problems involves the basic role of a university. In essence, a university's main objectives must be to educate its students, to generate new knowledge, and to preserve and disseminate existing knowledge. To take just one aspect of this charter, faculty members are expected to publish the results of their work in the open literature, and in science this requirement of publication frequently takes on a note of urgency. Yet industry often has a vested interest in keeping the results of its research secret, to protect information that might give it an advantage over a competitor. Thus, agreements whereby a university conducts research for a company must consider the university's responsibility to maintain the free flow of information.

Another potential problem involves the granting of patent rights. A 1980 law gave universities the right to retain patents arising from research sponsored by federal funds, and some universities have used this provision to generate revenue (for instance, Stanford University has earned several million dollars from its patent on the basic process used in making recombinant plasmids). Generally, universities also retain the patent rights on research supported by industry. But the nature of the licenses transferring property rights from universities to companies, especially their exclusivity and royalty provisions, can generate controversy. Problems may also arise with researchers conducting patentable work who are involved both with a university and with a private company.

Such situations are representative of other conflicts of interest that may result from university-industry alliances. For instance, a faculty member's involvement with a private company may affect his or her relationships with students and other faculty. Industrial support for academic research may also reorient that research away from the scientific interest inherent in a project and toward more commercial considerations.

In the early 1980s, when a number of university-industry agreements were being established in the life sciences, these and other questions were extensively discussed in several national forums. More recently, the discussion has been shifting to the local level as universities and companies have gained experience with the first wave of agreements. The concerns, too, have been changing, as the institutions involved have resolved the most obvious difficulties. The questions now coming to the fore have a more individualized cast. How effective have past arrangements been in meeting the needs of each institution? Has the administration of these alliances suffered from any obvious flaws? How have these arrangements been affecting the responsibilities and characteristics of each institution? Should the federal government or state governments take additional steps to encourage the formation of university-industry agreements?

Such questions are best answered in the context of specific instances of university-industry cooperation. Among the most notable of these instances has been the program set up in St. Louis by Monsanto and Washington University.

The Monsanto-Washington University Agreement

The negotiations between Monsanto and Washington University to establish a cooperative research program were in progress at the height of public interest in university-industry relations. As a result, according to David M. Kipnis of the Washington University School of Medicine, the agreement can be seen as a sort of test of "whether a partnership between two very different institutions is possible, and whether the separate though overlapping interests of both can be furthered without compromising the principles and values of each."

Both institutions brought definite goals to the negotiations.

Monsanto was interested in developing new products and in strengthening another St. Louis institution with which it has had many associations. The university was interested in extending its long-term commitment to basic biomedical research and in benefiting the St. Louis economy and community. Both viewed their geographic proximity as a significant advantage, since it would facilitate exchange and communication.

The university recognized from the beginning, according to Kipnis, that its foremost asset was its nationally known scientists. Thus, the first requirement of any association was that it preserve the scientific strength and reputation of the university. "It was clear that any significant sacrifice of scientific integrity and independence for short-term financial gain might very well result in serious long-term consequences," says Kipnis. "Another way to put it is simply that Washington University did not want to lose the very qualities that made it an attractive partner for Monsanto in the first place."

In setting up the agreement, the university was concerned about many of the generic issues that have surrounded university-industry relations. "Would the university be perceived as potentially diverting scientists from their primary areas of concern in traditional academic environments?" asks Kipnis. "Would the agreement influence graduate education by focusing on potential commercial ventures? Would the issues of secrecy and confidentiality thwart or interfere with the free interchange of scientific information, which is absolutely fundamental for scientific progress? Would a few academic stars become further enriched at the expense of other members of the faculty, both in terms of time and resources?" An important aspect of the negotiations, according to Kipnis, was that Monsanto's management fully understood and accepted these concerns.

The negotiations went on for about a year. The first few months were spent developing guidelines for the program to address the concerns of the university and to ensure that the venture would be profitable for the company. A retreat was then held in which scientists and managers from the university and the company discussed the state of the art, the future of the field, and their own goals and ambitions.

Washington University then submitted a proposal to Monsanto. "It was mutually decided that the program would focus on one defined area of biological science, broad enough in scope to take advantage of all of the perceived opportunities that many of us wished to pursue, focusing on a university strength and an area where we wished to broaden our own research, and perceived as an exciting area for industry," says Kipnis. "Therefore, the theme of the program was established as proteins and peptides regulating cell-cell communication and function."

A legal document of about 50 pages was prepared, signed, and distributed to newspapers and to Congress as a public document. It called for a five-year agreement that is renewable every three years. It is an agreement only between the two institutions, with no one investigator being singled out for special treatment. "By being prevented from benefiting as individuals from any commercial results of their effort, the faculty will not be tempted to turn away from its academic commitments and priorities," says Kipnis. "A full-time faculty ought to spend its full time in its academic pursuits. If one wishes to engage in commercial pursuits, then one has to do that full time also." Funds from the program go directly to the medical school. They are then allocated by an advisory committee composed of five scientists and managers from Monsanto and five scientists and administrators from the university.

It was decided during the negotiations to keep the funds from Monsanto at less than 10 percent of the medical school's research budget. The total budget of the Washington University School of Medicine is currently about $100 million per year, of which about $55 million comes from NIH, making the school one of the top three or four research centers in the country in terms of NIH funding. Thus the initial agreement involved a total grant of about $6 million per year, and the agreement has subsequently been expanded to reach $8.5 million to $9 million by 1987.

The entire faculty is eligible to participate in the program. Members are asked to submit letters of intent with a brief description of the research they propose to do. Out of 40 to 50 letters received annually, the advisory committee has been approving somewhere between six and nine projects, depending on the funds available. The projects have received grants of between $40,000 and $500,000 and have extended for periods ranging from a year and a half to three years. Currently, 38 investigators are being funded, spread across the entire hierarchy of the faculty, in a total of 21 different research projects. Twelve of the medical school's 19 departments have investigators receiving funds from the program.

Two kinds of research projects are being funded: exploratory and specialty. "Exploratory projects deal with fundamental research on basic scientific questions, with a focus on proteins or peptides," explains Kipnis. "Examples of this kind of research are studies of

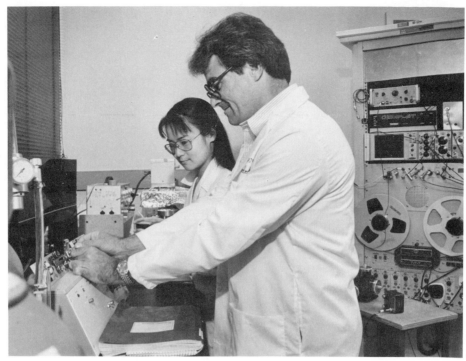

Washington University

Researchers at Washington University work on specific projects funded through a cooperative agreement between the university and Monsanto. In developing the agreement, both institutions had to deal with many of the concerns that have marked university-industry relations.

glycosylation mechanisms, the regulation of gene expression, the biology of receptors, et cetera." Specialty projects focus on "specific proteins and peptides that modulate cellular function and in which is seen the possibility of commercial utility in terms of technologies or products. Examples of research that are well known are atrial peptides, research on mediators of immune reactions, the structure function of proteins regulating coagulation or thrombolysis, or the protein products influencing oncogene function."

In addition to the oversight of the advisory committee, the agreement calls for external peer review on a regular basis. Every three years an independent group of internationally recognized scientists will be brought to Washington University and asked to review the quality of the research and the impact of the program on the university and company. The first of these reviews was held in October 1985.

According to the agreement, any patents that originate from the program will reside with the university, with Monsanto agreeing to provide legal support in applying for patents. In turn, the university investigators have agreed to keep confidential any information given to them by Monsanto that is corporate property. Corporate scientists have likewise agreed to maintain the confidentiality of any material from the university until it is published or presented in public. There is a 30-day delay during which abstracts and papers can be reviewed by patent attorneys for any material that may be patentable. "But our recent experience has been that this has not, in essence, hindered publication," says Kipnis. Similarly, any royalties deriving from the program will be returned to the medical school, to be divided among the medical school and the department and laboratory that conducted the royalty-generating research, just as the school now handles professional fees.

The agreement took effect in September 1982, and according to Kipnis, "the success of the program has exceeded our expectations. The interaction of scientific groups of high quality has gone on without the need to force it. It has been a natural evolution of true scientific curiosity. . . . We have had, in essence, no complaints, even by those whose grants have been disapproved."

The university has garnered a number of benefits from the program, Kipnis points out. The availability of funds has allowed the university to move rapidly into areas of interest, often more quickly than would be possible through the cumbersome and time-consuming process of receiving federal grants. "It has allowed us a certain degree of flexibility," he says, "where investigators coming up with unique observations or unique findings can come to the committee and ask for supplemental funds or, indeed, during the year, for the initiation of new projects." The program has also allowed young investigators to do innovative work that they would probably be unable to do through conventional channels. "Availability of funds has allowed us a certain entrepreneurial spirit, not in terms of commercial gains, but in terms of scientific enthusiasm," remarks Kipnis. "That has been very helpful." The two institutions have even established a monthly seminar program, given at alternate locations, and an annual retreat on a subject of mutual scientific interest.

In general, concludes Kipnis, "we are extremely pleased and optimistic on the basis of our initial findings, and we believe it is a desirable experiment and one that may open up new possibilities for both society at large as well as for industries and universities."

Additional Readings

Wil Lepkowski. 1984. "University/Industry Research Ties Still Viewed with Concern." *Chemical and Engineering News* (June 25):7-11.

National Science Board. 1982. *University-Industry Research Relationships: Myths, Realities and Potentials.* Washington, D.C.: National Science Foundation.

Robert D. Varrin and Diane S. Kukich. 1985. "Guidelines for Industry-Sponsored Research at Universities." *Science* 227(January 25):385-388.

10

Biotechnology in Japan: A Challenge to U.S. Leadership?

ALTHOUGH FEW OF ITS PRODUCTS have yet reached the market-place, biotechnology clearly has the potential to become a major commercial enterprise. As such, it could eventually join the handful of advanced technologies that have assumed a prominent role in the U.S. economy. These technologies have formed the basis of core industries whose potential for increased productivity and falling costs has contributed to economic growth and rising employment. Their effects have also radiated into other areas of the economy, even into traditionally low-technology or service sectors.

In addition to their effect on the domestic economy, high-technology industries have a critical influence on the U.S. balance of trade. The commercial products of advanced technologies have in the past shown a positive trade balance for the United States, while other manufac-tured goods have shown a trade deficit. The United States currently holds the highest market share of the industrialized countries' exports of high-technology products, but that share has been declining in recent decades. At a time of high U.S. trade deficits, the balance of trade in high-technology products inevitably draws special attention.

The United States currently enjoys a sizable lead in transforming the

This chapter includes material from the presentations by Gary R. Saxonhouse and Irving S. Johnson at the symposium.

SCIENTISTS GET FIRST CLOSE LOOK AT A MOLECULE

results of basic biomedical research into commercial products. However, the economic appeal of biotechnology as an expanding high-technology industry has not escaped the attention of other industrialized countries. "Most other developed countries have targeted biotechnology as a goal," says Irving S. Johnson of Eli Lilly. "These countries include the United Kingdom, France, West Germany, Russia, and, most vigorously, Japan. The national efforts of these countries have ranged from modification of guidelines for carrying out research, to legislative assistance, to financial support of private companies, to dismantling of unneeded antitrust legislation."

Unlike other national governments, the federal government of the United States has no explicit policies to encourage the development of biotechnology. As a result, Johnson feels that "the United States may well be flirting with the loss of its current, but in my view fragile, competitive lead."

The country expected to be the United States' leading competitor in commercializing genetic engineering is Japan. A major contributor to the strength of the Japanese effort has been the Japanese government's promotion of the field. "It is very clear that the Japanese government is interested in biotechnology at the highest levels," says Gary R. Saxonhouse of the University of Michigan. "At the 1983 economic summit held in Williamsburg, Virginia, Prime Minister Nakasone astounded the other participants, and particularly, I am told, President Reagan, by spending as much as 15 or 20 minutes attempting to enlighten the President on the nature of recombinant DNA and its prospects for the future, an example of a strategy for industrial development in both Japan and the United States that Prime Minister Nakasone alleged would lead to more harmonious economic relations."

But even though the Japanese interest in biotechnology is strong, its origins are relatively recent, according to Saxonhouse. Not until 1980, with the success of Genentech's public stock offering and the Supreme Court's decision allowing the patenting of microorganisms, did the Japanese government begin to expand its previously low-key efforts in biotechnology. Ironically, says Saxonhouse, the Japanese feared that "the American government was weaving around the biotechnology industry a network of protective patents, and that, in some fashion, Japanese firms were going to be denied access to this important technology. . . . Their interest was a reaction to developments in the United States."

Nevertheless, the Japanese have quickly built a strong program in biotechnology. The government has supported biotechnology research not only in universities but in government institutes and selected industries. It has helped form consortia of industries to coordinate research and development in biotechnology and reduce duplication of effort. "The Japanese have frankly admitted that, whereas they may be five years behind in biotechnology, they intend to make up the difference quickly, by scrapping older technology and improving on new technology imported from the United States," says Johnson. "That has a familiar ring to me."

National Policies Affecting the Commercialization of Biotechnology

Although the U.S. government has no explicit policies concerning biotechnology, federal policies have exerted, and will continue to exert, an important indirect effect on the field, both in the research laboratories where it was developed and in the firms where it is being commercialized. These policies and their counterparts in other countries will have an important influence on the competitiveness of biotechnology firms in world markets.

Governmental Funding of Basic and Applied Research

The U.S. government has concentrated its direct support for biotechnology on basic scientific research conducted in universities and other research institutes. The main source of this support has been the National Institutes of Health, with lesser amounts provided by the National Science Foundation, the U.S. Department of Agriculture, the Department of Energy, the Department of Defense, and the Agency for International Development. All told, this federal support of basic research in biotechnology amounts to some $750 million per year.

It is difficult to estimate just how much other governments are spending on biotechnology, partly because of problems in defining the term and partly because of the inevitable overlap between biotechnology and other scientific and industrial endeavors. But after a careful review of the available statistics, Saxonhouse concludes that the Japanese government is probably spending no more than $60 million per year on biotechnology research. Thus, he points out, "the American government probably sponsors and conducts 10 times as much activity, at a minimum, as the Japanese government does."

One difference between government-sponsored research in Japan and in the United States involves the use of funds. Whereas U.S. funding goes almost entirely for basic research, Japanese funding is directed much more heavily toward more applied research, including work on product development and scale-up. For instance, the Ministry of International Trade and Industry (MITI), under its Office of Biotechnology Promotion, is sponsoring research in three broad areas: recombinant DNA technologies, large-scale cell cultures, and bioprocess engineering. Part of its funds go directly to a group of 14 chemical and energy companies, which are working together on projects in each of the three research areas MITI has selected.

MITI is not the only Japanese agency that supports research in biotechnology. For instance, the Ministry of Agriculture, Forestry and Fisheries provides much of the Japanese government's support for cell fusion techniques. But this diversification of effort can have disadvantages as well as advantages, according to Saxonhouse. "Biotechnology is not easily incorporated within the purview of any particular Japanese government ministry. The coordinating apparatus among these government agencies looks no better than the coordinating apparatus that you would probably find in the United States among the many different government agencies here that have an interest in biotechnology research."

Also, Saxonhouse believes that governmental support for biotechnology in Japan has had a narrower focus than might be expected. The bulk of the funds has gone toward commercial sectors that are structurally depressed, including agriculture, energy, chemicals, pulp and paper, and textiles. "The interest in biotechnology in Japan, in particular MITI's interest, is largely centered on helping companies that are presently in difficulty," says Saxonhouse. "This is a very different kind of policy than the Japanese government pursued in the semiconductor industry."

It now appears as if the first test of international competitiveness in biotechnology will be in the pharmaceuticals industry. The Japanese pharmaceuticals industry has been trying to increase its global market share in recent years, and the Japanese government has been encouraging this expansion through such actions as changes of patent laws and pricing guidelines in the national health insurance system. Another highly competitive sector in the early years of biotechnology will probably be the specialty chemicals industry, where Japan already has a substantial market presence.

The Financing of the Biotechnology Industry

The single most important factor contributing to the United States' current lead in biotechnology has been the establishment and growth of the new biotechnology companies. Bolstered by ready supplies of venture capital and by tax and investment laws encouraging their development, these firms have demonstrated to industries around the world the commercial viability of the field.

With a handful of exceptions in the European Economic Community, there are no start-up biotechnology firms outside the United States. In Japan there are more than 200 firms working on the commercial applications of biotechnology, but they are all established firms from traditional industrial sectors. "Start-up firms are encouraged in the United States through the tax code in a way that start-up firms simply are not encouraged in Japan," notes Saxonhouse.

Firms working on biotechnology in Japan generally rely on internal sources of funds and on bank loans to finance their research and development. Public stock offerings, venture capital, and related means of equity financing are of relatively minor importance in Japan. To some extent, the promotion by the Japanese government acts as a signal to the financial system to be more receptive to requests for funds to finance biotechnology research and development, according to Saxonhouse. But, with the single exception of a recently enacted 7 percent tax credit on investment in R&D-related equipment, there are

no provisions within the Japanese tax code that are any more benefi-
cial to biotechnology than to other industries, nor have any loans
been given to Japanese companies working in biotechnology at low
or concessional rates. To the extent that the industrial policies developed by the Japanese government do boost the development of biotechnology, says Saxonhouse, they can be seen more as substitutes
for the efficient capital markets of other countries than as independent influences.

The Regulation of Biotechnology

Many foreign countries, including Japan, generally followed the
United States' lead in first establishing guidelines on recombinant
DNA research and then gradually easing them as the initial fears
proved groundless. However, the guidelines in Japan remain significantly stricter than in the United States, which may prevent some
promising research from being done there.

Japan also has the most restrictive regulations concerning health
and safety for new drugs, biologics, and medical devices of any of the
countries that are commercializing biotechnology. In the past Japan
has used these regulations, which include approval policies, product
standards, and testing procedures, as nontariff barriers to the import
of pharmaceuticals. In 1983 the laws were changed to give equal
treatment in principle to foreign products, and "there have been a
number of anticipatory steps within the product approval machinery
within Japan's Ministry of Health and Welfare to ensure that this
[discriminatory approval] does not happen," says Saxonhouse. But as
in other areas of dispute involving U.S.-Japanese trade, the effects of
old laws still linger.

Multilateral trade agreements like the General Agreement on Tariffs and Trade work to lower tariffs, discourage nontariff barriers to
trade, and eliminate governmental subsidies to industry. U.S. trade
law also provides American companies with ways to seek relief from
unfair import or export practices. For instance, section 301 of the Trade
Act of 1974 allows firms to petition the U.S. government to enforce
their rights under trade agreements or to negotiate to eliminate
actions by foreign governments that limit their access to foreign
markets.

The Protection of Intellectual Property

The degree of protection that a company can obtain over products or
processes it has developed can be an important factor in determining

its competitiveness. In the United States, inventors can apply for a patent up to a year after an invention is described in a scientific journal or meeting. Patent applications can also be kept secret until a decision on the application is made. In Japan, the grace period for applying for a patent after public release of the patentable information is only six months. About 18 months after a patent application is filed, the information is released to the public—even if the patent has not yet been issued—precluding the option of trade secrecy once the decision is made to pursue a patent.

Partly as a result of the Supreme Court's 1980 decision in *Diamond* v. *Chakrabarty,* the range of patentable subject matter in the United States is very broad. Japan, in contrast, does not grant patents on medical processes that involve the human body as an indispensable element. Japan's strict health and safety guidelines regarding genetic engineering may also restrict the patenting of organisms viewed as hazardous.

The Availability of Trained Personnel

The biotechnology industry has already created some 5,000 to 10,000 jobs in the United States. At this early stage in the industry's development, these jobs are predominantly for highly trained research scientists, such as molecular biologists, immunologists, and related technologists. As biotechnology moves toward large-scale manufacturing of its products, the personnel needs of many biotechnology firms will shift. To design and develop the production technologies needed for manufacturing, these firms will need more process-oriented researchers, including bioprocess engineers and industrial microbiologists.

In Japan a sharp division between basic and applied research in universities, along with limited support for basic scientific research, initially caused a shortage of experts trained in the basic techniques of genetic engineering. The Japanese are taking a number of steps to ease this shortage. For one, they are drawing on their extensive historical experience with fermentation techniques in developing production methods in biotechnology.

The Japanese are also sending researchers abroad to study. Many Japanese scientists in disciplines related to biotechnology are currently working and studying in the United States. For instance, more than 200 are currently working at NIH, a number that is bolstered by an accounting provision that allows laboratories to exceed their maximum staff sizes with foreign nationals in temporary positions. At the same time, the number of American scientists and engineers traveling abroad to study has been falling steadily, despite the fact that there are a number of eminent foreign institutes in the field, such as Japan's Fermentation Research Institute, where American researchers could receive valuable training.

Finally, Japanese government and industry are attempting to induce Japanese nationals working abroad to return to the country and are retraining scientists and technicians within Japan. Retraining of industrial personnel is much more common in Japan than it is in the United States, and Japan's extraordinary ability to overcome weaknesses in its labor force is one of its great strengths. This ability is also one of the factors that have enabled Japan to quickly become a global competitor in biotechnology.

International Technology Transfer

The imbalance between foreign researchers studying in this country and American researchers studying abroad is one way in which technology moves out of the United States. Another is joint ventures between American and foreign biotechnology firms. Japanese companies, in particular, have supported a large amount of contract research by American biotechnology firms, enabling the Japanese companies to keep up with the state of the art in biotechnology. With biotechnology still in a knowledge-intensive phase, it is possible that the movement of this information could help foreign firms establish themselves more favorably in world markets, to the detriment of U.S. firms.

The openness of the American university system and the frequent movement of personnel within American industry contribute to the diffusion of information in science and technology, both domestically and overseas. In Japan, however, researchers usually stay with a single firm or university throughout their lives. This results in much less communication and cooperation among scientists and engineers in Japan, which in turn tends to inhibit the flow of technological information out of the country. Indeed, the joint R&D programs sponsored by the Japanese government are to some degree an attempt to overcome this intranational insularity.

The differences in the openness and degree of communication among researchers in the United States and researchers in Japan may partly explain how the Japanese have managed to build a strong effort in biotechnology so quickly. The results of American research are available to all through open publication, while Japanese research is much harder for Americans to receive and use. "If we look at what the Japanese government spends for biotechnology, chances are that most of it is of use almost entirely to the Japanese biotechnology industry, and of relatively small benefit to foreign biotechnology firms," Saxonhouse points out. "On the other hand, much of what the U.S. government spends is as useful to Japanese biotechnology firms as it is to American biotechnology firms."

It must be remembered that foreign technology does flow into the United States, although it is difficult to assess the magnitude of this flow. The most notable example in biotechnology is the process for making monoclonal antibodies, which was developed in the United Kingdom. High-quality research in molecular biology, immunology, and bioprocess engineering is also being conducted and published in other countries. But most observers would agree that the net flow of technology transfer in biotechnology is currently out of the United States. It is too soon to tell if this will significantly impair the competitiveness of U.S. biotechnology firms.

Possible Governmental Responses to International Competition in Biotechnology

The U.S. government would have a number of alternatives if it were to decide that biotechnology is important enough to the future of the nation's economy to warrant direct governmental assistance. At one extreme, it could adopt some of the more overt targeting practices of foreign countries, including direct development aid to private companies, industrywide assistance through low-interest loans or tax credits, or interagency oversight to coordinate federal policies and industrial R&D. However, it is highly unlikely that many of these options would be accepted in the United States, given the traditional roles of government and industry.

The federal government could also act to boost the competitiveness of U.S. biotechnology firms in a number of indirect or less industry-specific ways. According to Johnson, such actions together could provide the United States with a much more consistent and effective approach to promoting biotechnology than now exists. Among the steps he suggests as part of such a policy are the following:

- Further increase federal support of basic research related to biotechnology, particularly in agriculture.
- Target federal assistance to bioprocessing and applied microbiology centers, possibly by funding through universities.
- Reassign some of the U.S. fellowships used to train foreign scientists at leading biotechnology centers in this country to the training of American scientists at foreign technology centers.

• Strengthen intellectual property law through the formation of a scientific advisory committee in biotechnology in the Patent and Trademark Office.

• Export products rather than technology whenever possible, and obtain adequate returns from the export of technology should such export be necessary.

• Clarify and update the tax code to provide incentives to conduct research and development in biotechnology.

• Reexamine antitrust regulation to further cooperation among companies conducting basic research in biotechnology.

Additional Readings

Mark D. Dibner. 1985. "Biotechnology in Pharmaceuticals: The Japanese Challenge." *Science* 229(September 20):1230-1235.

Ralph W. F. Hardy. 1985. "Biotechnology: Status, Forecast, and Issues." Pp. 191-226 in *Technological Frontiers and Foreign Relations*. Washington, D.C.: National Academy Press.

National Research Council, Office of International Affairs, Panel on Advanced Technology Competition and the Industrialized Allies. 1983. *International Competition in Advanced Technologies: Decisions for America*. Washington, D.C.: National Academy Press.

Office of Technology Assessment. 1984. *Commercial Biotechnology: An International Analysis*. Washington, D.C.: U.S. Government Printing Office.

U.S. Department of Commerce, International Trade Administration. 1984. *High Technology Industries: Profiles and Outlooks—Biotechnology*. Washington, D.C.: U.S. Government Printing Office.

Index

A

Academia, *see* Universities
Acquired immune deficiency syndrome (AIDS), 3, 20, 28
Adenosine deaminase (ADA), 45-48
Adhesives, 15, 21
Agency for International Development, 79, 110
Agriculture, *see* Genetic engineering, agricultural; *specific crops and plants*
Agrobacterium tumefaciens, 35, 36, 40
AIDS, *see* Acquired immune deficiency syndrome
Alexander, Martin, 54, 55-60
Amino acids, 5, 15, 18, 20, 37, 94
    See also Proteins
Anderson, W. French, 43, 45-48, 51-52
Animals, *see* Genetic engineering, animals; *specific animals*
Antibiotics, 1, 5, 14, 21, 30, 35, 95
Antibodies, 5, 14, 20, 73
    See also Monoclonal antibodies
Antigens, 25-26
Asilomar conference, *see* International Conference on Recombinant DNA Molecules
Atherosclerosis, 52, 89

B

B lymphocytes, 25, 27
Bacillus thuringiensis, 34
Bacteria, 1, 3, 11, 15-17, 20-21, 24, 28, 32, 33, 59, 66, 71, 95
    See also specific bacteria
Bentley, Orville G., 64, 78
Beta-thalassemia, 43, 45
Biological synthesis, 14
Biologics, 13, 72-75, 78, 82, 115
Biotechnology firms
    challenges to, 89-90
    characteristics, 85-87
    funding, 39, 85-87, 111-112
    importance of in United States, 111
    successful, 87-89
    types, 10, 12-13, 84-85
    See also Industry
Bone marrow cells, 45-48

C

Cabinet Council on Natural Resources and the Environment, 79
Cabinet Council Working Group on Biotechnology, 49, 79-81
Cancer, 19, 28, 89
Cattle, 30, 31, 40, 41
Cells
    bone marrow, 45-48
    cloning, 3, 5, 27
    culture techniques, 34
    fermentation, 5, 22-24
    fusion, 25, 75
    germline, 7, 49-51, 52
    growth, 24

hosts, 17, 34-35
molecular machinery, 15-18, 19, 22
sexual reproduction, 6-7
somatic, 7, 44-50
transformed, 47
tumor, 28, 29, 36
university-industry research, 103, 104
    See also DNA; Genes; Human gene therapy; Recombinant DNA
Centers for Disease Control, 79
Centocor, Inc., 87-90
Cetus, 86, 90
Chemicals, 4, 5, 9, 16, 21-22, 28, 76-77
Chemicals industry, 10, 14, 15, 21-22, 84, 85, 111
Chloroplasts, 35
Cholesterol, 52
Chromosomes, 16, 35, 41, 50, 51
Cline, Martin, 43, 44
Cloning, 3, 5, 27
Colibacillosis vaccine, 31
Commercialization of biotechnology, national policies affecting, 109-115
Congressional role in biotechnology, 77, 81-83
    See also U.S. government
Consulting arrangements between universities and industry, 99-100
Contracts between universities and industry, 99-100
Corn, 37, 56, 71
Corynebacterium, 21
Costle, Douglas, 2
Cosmids, 16
Crops, *see specific crops*
Cunningham, Brian, 64, 74-75
Cystic fibrosis, 45

D

Deoxyribonucleic acid, *see* DNA
Department of Agriculture (USDA), 7, 9, 66, 70, 77, 78, 80, 110
Department of Commerce, 79
Department of Defense, 79, 110
Department of Energy, 79, 110
Department of the Interior, 77, 79
Department of State, 79
Diabetes, 18
Diamond v. Chakrabarty, 91-92, 113
Diseases, 20, 28, 31, 40, 43, 45-47, 59, 85, 89
    See also specific diseases
Dispersal of genetically engineered organisms, 59-60
DNA
    extra, 59
    foreign, 38, 39-40, 49-51, 61
    human gene therapy, 46-50
    probes, 95
    transferrence, 8, 15-18, 34-42, 56, 60
    See also Recombinant DNA
Drugs, 3, 6, 13, 18, 19, 21, 30, 72-75, 82, 112
Dyes, 15, 21